The World Through Your Nose: Unveiling the Secrets of Smell

Sartre

TABLE OF CONTENTS

CHAPTER 1. INTRODUCTION

Transduction of stimuli from the external environment is achieved through our sensory systems and converted to an electrical signal. The environmental stimuli that are detected by the olfactory system are volatile elements in the air, commonly referred to as odors. Odors are detected by neurons in the olfactory epithelium (OE) that line the nasal cavity. After an odor is detected, the information is transferred from the OE to the olfactory bulb (OB) in the central nervous system and then subsequently relayed to higher areas of the brain (DeMaria and Ngai 2010).

The olfactory system has a unique capability for regenerating neurons. The olfactory sensory neurons (OSNs) that are responsible for the perception of odorant molecules are constantly dying. It is generally accepted that the high death rate of OSNs is caused by the exposure of OSNs to the external environment, which is necessary for the detection of odorant molecules in the air. This makes OSNs unique in that they are the only neurons in the body that are exposed to the external environment. However, this additionally exposes the OSNs to toxic elements in the air (e.g. pollution) that can lead to the death of the neuron. Maintenance of the sense of smell is achieved by replacing the ever dying OSNs with new neurons derived from the stem cells located at the base of the OE.

In order to better understand the regenerative process of the olfactory system, this introduction will first describe the organization of the olfactory system, then we will introduce the concept of axon guidance and finally, we will describe the process of regeneration in the olfactory system both during the events of normal turnover and after injury.

Organization of the Olfactory System

Aside from the main olfactory system, that will be the focus of this review, the accessory olfactory system is also located in the nasal cavity. The accessory olfactory system is comprised of the vomeronasal organ and nerve, and the accessory olfactory bulb. For additional information on this system, the reader is referred to a recent review by Mohrhardt et al (2018). Furthermore, there are two additional systems associated with the detection of odors that include structures such as the septal organ (organ of Masera) and the Grüeneberg ganglion that will not be discussed here but the reader is again referred to a review of these system by Minghong Ma (2010). The main olfactory system can be divided into two primary components: the main OE and the main OB, which are the most relevant components to the discussion of OSN regeneration.

Olfactory Epithelium

The OE together with its underlying connective tissue, the lamina propria, compose the olfactory mucosa. The olfactory mucosa lines the dorsal and caudal portions of the nasal cavity. The nasal septum that divides both sides of the nasal cavity, along with the turbinates, delineate the shape of the nasal cavity. The OE is a pseudostratified epithelium and is, broadly speaking, comprised of three cell types: OSNs, supporting cells, and basal cells (Farbman 1992). Of these cell types, OSNs represent the larger population of cells in the OE and are responsible for conduction of sensory information from the OE to the OB. However, the other cell types also play an important role in olfaction.

10

OSNs are bipolar neurons that have a cell body contained in the OE and extend an apical dendrite toward the nasal cavity that ends in a knob from which multiple non-motile cilia arise. On the opposite side of the OSN cell body, an axon projects into the lamina propria. The axons of the OSNs converge to form (in humans) cranial nerve I and later cross the most anterior-basal portion of the skull called the cribriform plate to enter the OB. After entering the OB, the axons defasciculate and make specific synaptic contacts (Farbman 1992). The OSNs, as previously discussed, are responsible for the detection of the volatile odorants in the air and do so by the binding of the odorants to G-protein coupled receptors known as odorant receptors. While the odorant receptors are located throughout the entire OSN, receptors expressed on the olfactory cilia are associated with odor detection. The olfactory cilia extend laterally in the mucosal layer of the nasal cavity, which allows for larger surface area for the binding of odorants. The binding of the odorant to the odorant receptor leads to the depolarization of the neuron and creation of an action potential that travels from the dendrite to the cell body of the neuron and down the axon to the OB.

Supporting cells can be divided into two biochemically separate cell types: sustentacular cells and microvillar cells (Schwob and Costanzo 2010). One of the functional roles of supporting cells includes neuroprotection through detoxification via the expression of the enzymes cytochrome P450 and Glutathione S-transferase (Gu et al. 1998; Ling et al. 2004; Whitby-Logan et al. 2004). Additionally, supporting cells have been shown to play a role in phagocytosis of dead OSN cell bodies in a bulbectomy injury model (Suzuki et al. 1996).

Basal cells are a group of stem cells that lie on the basal membrane of the OE

near the juncture of the epithelium and lamina propria. The basal cells can be divided

into two distinct cell types: horizontal basal cells and globose basal cells (Farbman

1992). The differing roles of these two cell types will be discussed further when

addressing the process of regeneration of OSNs.

Olfactory Bulb

The OB is the structure in the CNS in which the axons of the OSNs from the OE

make synaptic contact. The OB is comprised of six distinct layers that contain unique

cell types and/or synaptic networks (*Figure 1.1*). The first and outermost layer of the OB

is the olfactory nerve layer (ONL), which is primarily composed of axons of OSNs

coming from the OE. In addition, a specialized glial cell type called the olfactory

ensheathing cell (OEC) is present in the ONL and plays a role in the process of axon

guidance. The second layer from the periphery of the OB is the glomerular layer (GL),

which is characterized by circular structures called glomeruli in which the OSNs make

synaptic contact with the projection neurons of the OB. Delineating the glomeruli are a

population of interneurons called periglomerular cells that have dendritic projections that

arborize within a glomerulus. Periglomerular neurons are interneurons that receive

synapses from the OSN axons and establish dendrodendritic synapses with the

projection neurons. The third layer of the OB is called the external plexiform layer (EPL),

where there is a network of dendrodendritic synapses between projection neurons and

granule cells, a second type of interneurons of the OB. In addition, the EPL contains the

cell bodies of tufted cells, one of the two types of projection neurons in the OB. The

fourth layer of the OB is called the mitral cell layer (MCL), which contains the cell bodies of the mitral cells, the OB's second type of projection neuron. The fifth layer of the OB is

the internal plexiform layer (IPL), which contains axons from the projection neurons that

travel through the OB and leave to form the lateral olfactory tract and subsequently

contact higher cortical targets. The sixth layer of the OB is the granule cell layer (GCL)

and contains the cell bodies of the granule cells mentioned earlier (Greer et al. 2008).

Additionally, deep in the GCL is the rostral migratory stream (RMS), a mass of newly

born migrating neurons born in the subependymal zone of the lateral ventricle. After

entering the OB, they detach from the RMS and migrate laterally within the OB and

differentiate into new interneurons (both periglomerular and granule cells). The process

and function of the RMS will not be discussed in this review, but the reader is referred to

reviews that discuss the RMS in great detail (Whitman and Greer 2009; Hardy and

Saghatelyan 2017).

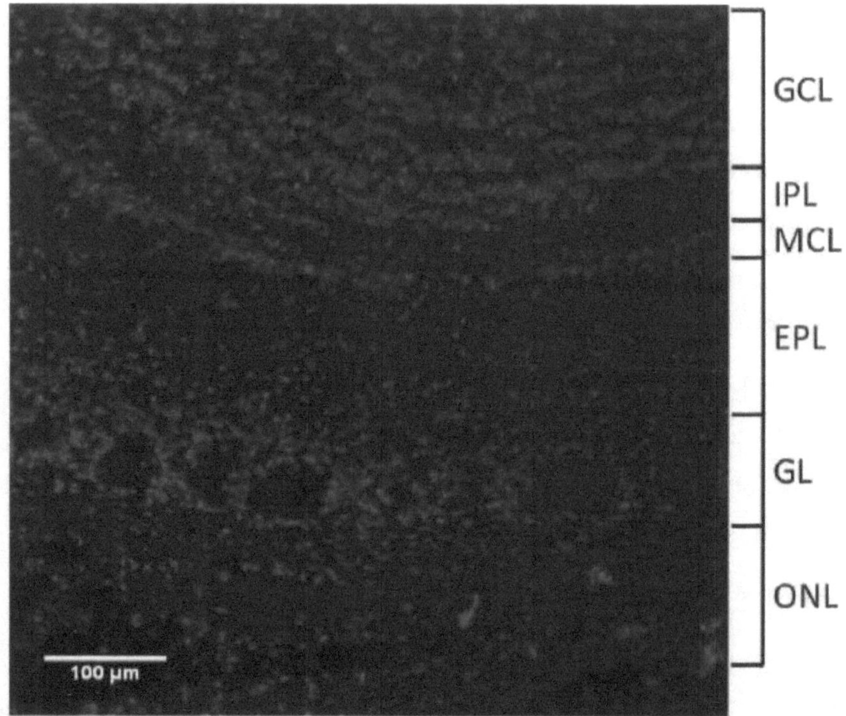

Figure 1.1. DAPI staining showing the different layers of the OB. ONL- olfactory nerve layer; GL- glomerular layer; EPL- external plexiform layer; MCL- mitral cell layer; IPL- internal plexiform layer; GCL- granule cell layer.

Axon Guidance

The process by which OSN axons extend to their specific synaptic targets within the OB is known as axon guidance and targeting. This extremely precise process has been extensively studied but is still not well understood due to the extreme complexity of this process. To understand the complexity that underlies OSN axon extension and targeting to a specific synaptic target, one must consider the following list of factors.

14

Each OSN expresses only 1 allele (monogenic and monoallelic expression) out of

approximately 1,200 odorant receptors genes, which in mice theoretically results in up

to 2,400 molecularly distinct sub-populations of OSNs (Chess et al. 1994; Serizawa et

al. 2003). OSNs that express the same odorant receptor are intermingled in the OE with

OSNs that express different odorant receptors. Yet, all of the OSN axons expressing the

same odorant receptor converge into 2-3 glomeruli per olfactory bulb (Richard et al.

2010) and each glomerulus is innervated by only axons expressing the same odorant

receptor (Treloar et al. 2002). Finally, OSN axons are randomly mixed as they navigate

from the OE to the OB and convergence of axons expressing the same odorant

receptor seems to occur proximal to their targeted glomerulus (Ressler et al. 1993;

Miyamichi et al. 2005). With these factors considered, the precision by which axons

extend and reach their synaptic target is truly remarkable. Many molecules and

signaling pathways have been identified that seem to play a role in the extension of

OSN axons and the guidance toward their synaptic targets. The following contains a

brief exploration of a few of the factors that have been implicated in the process of axon

guidance and for more detail on axon guidance molecules in the olfactory system, the

reader is referred to other excellent reviews (Treloar et al. 2001; Mombaerts 2006;

Takeuchi and Sakano 2014).

Epithelial Zone and Glomerular Position

Observations have long been made using retrograde tracing studies that suggest

that OSNs located in given areas of the OE project to a correlating area within the OB

(Astic and Saucier 1986; Pedersen et al. 1986; Schoenfeld et al. 1994; Levai et al.

2003). One study using Dil tracing demonstrated that the position of the glomeruli within the OB along the dorsal/ventral axis correlated with a dorsomedial/ventrolateral axis within the OE (Miyamichi et al. 2005). These observations suggested a zonal map within the OE that correlates with the glomerular map of the OB. The classic model of odorant receptor expression in the OE indicated that there were four distinct zones comprised of OSNs that expressed a limited number of odorant receptors largely unique from that of the other zones (Ressler et al. 1993; Vassar et al. 1993). However, it has since been shown that these zones overlap in odorant receptor expression (Norlin et al. 2001; Iwema et al. 2004; Miyamichi et al. 2005). While these zones may not have well defined borders with regards to odorant receptor expression, there have been correlations with specific zonal expression of different molecules implicated in axon guidance. The cell adhesion molecules RNCAM (neural cell adhesion molecule 2) and OCAM (olfactory cell adhesion molecule) have shown specific zonal expression in ventral zones 2-4 (Cremer et al. 1994; Alenius and Bohm 1997; Yoshihara et al. 1997). In accordance with this zonal distribution of axon guidance factors, it has recently been shown that transplantation of stem cells from the dorsal OE to the ventral OE leads to an upregulation of OCAM and downregulation of NQO1 (NAD(P)H Quinone Dehydrogenase 1), a dorsal OSN marker, suggesting that spatial cues present in the OE determine OSN identity and subsequently their glomerular target (Coleman et al. 2019).

Odorant Receptors

There is a general consensus that odorant receptors are necessary but not sufficient to fully account for the targeting of the OSN axons to their specific glomerulus (Wang et al. 1998; Feinstein et al. 2004; Feinstein and Mombaerts 2004). As an example of this, it has been shown that OSN axons in mice lacking functional odorant receptors still extend broadly to the correct area of the OB (Feinstein and Mombaerts 2004). This suggests that axonal navigation from the OE to the OB remains mostly unaffected by the odorant receptors. However, the function of odorant receptor expression seems to be more critical for the final targeting of a specific glomerulus within the OB. In support of this, we have recently shown that odorant receptor expression does not occur until after the olfactory sensory neuron axons have reached the OB (Rodriguez-Gil et al. 2015). Our findings suggest a model of OSN axon extension in two phases: an odorant receptor-independent phase from the OE to the OB and then an odorant receptor-dependent phase within to the OB to the specific glomerulus for each odorant receptor.

There are approximately 3600 glomeruli in each OB (Richard et al. 2010) and, as previously mentioned, this results in an average of 2-3 glomeruli in each OB for every odorant receptor gene. Glomerular location within the OB can be altered by even small changes in the gene encoding the odorant receptor and can shift targeting to adjacent/ectopic glomeruli (Gogos et al. 2000; Strotmann et al. 2000; Feinstein et al. 2004; Feinstein and Mombaerts 2004). Thus, the identity of the odorant receptor is key in determining the glomerular location within the olfactory bulb rather than an assigned location via other signaling mechanisms. However, it is noteworthy that this is not an inherent characteristic unique to odorant receptors. Replacement of the protein coding region of an odorant receptor locus with that of the β2 adrenergic receptor results in axons targeting a new glomerulus unique to this receptor (Feinstein et al. 2004). So, while the receptor expressed in the OSN axons are critical for determining the glomerular location within the OB, the mechanisms by which the receptor determines this location is still not well understood. Downstream signaling through the second messenger cAMP is important in determining the final position of the axon targeting (Imai et al. 2006; Chesler et al. 2007). Recently, however, phosphatidylethanolamine-binding protein 1 (PEBP1), a putative binding target for odorant receptors expressed in the axon terminal, is associated with formation of the OB's sensory map, and deletion of PEPB1 leads to an altered glomerular map (Zamparo et al. 2019). These findings could shed light on the mechanisms that underlie odorant receptor's role in axon guidance.

Other Molecules of Axon Guidance

In addition to odorant receptors, other mechanisms, including both attractive and repulsive cues, have been implicated in olfactory sensory neuron axon targeting (Treloar et al. 2001; Mombaerts 2006; Schwarting and Henion 2008). Several laboratories have described the spatio-temporal distribution of laminin, tenascin, semaphorins-neuropilins, cell surface markers and different carbohydrates, and correlated these observations with the formation and maturation of olfactory bulb glomeruli (Ring et al. 1997; Kafitz and Greer 1998; Crandall et al. 2000; Pays and Schwarting 2000; Williams-Hogarth et al. 2000; St John and Key 2001; Treloar et al. 2001; B.W. Lipscomb et al. 2002; B. Lipscomb et al. 2002; Walz et al. 2002; Henion et al. 2005; St John et al. 2006). Other candidate molecules characterized in the olfactory system include ephrins/Ephs, robos/slits, netrins/DCC, and Wnt/Frizzled (Astic et al. 2002; St. John et al. 2002; Marillat et al. 2002; Zaghetto et al. 2007; Booker-Dwyer et al. 2008; Rodriguez-Gil and Greer 2008; Wang et al. 2008). As an example, the dorsal/ventral axis navigation of the OB has been shown to be influenced by expression of Robo2/Slit1 (Cho et al. 2007; Cho et al. 2012) and Nrp2/Sema3F (Takahashi et al. 2010; Takeuchi et al. 2010). When examining the anterior/posterior glomerular location, Nrp1/PlxnA1 have been shown to have an inverse gradient from anterior to posterior and changes in their protein levels can alter the anterior/posterior location of a given glomeruli (Satoda et al. 1995; Imai et al. 2006; Nakashima et al. 2013). Nevertheless, the full repertoire of mechanisms that influence olfactory sensory neuron axon targeting remains topical and controversial (Mombaerts 2006; Schwarting and Henion 2008).

Cellular Contributions to Axon Guidance

An OSN axon interacts with several cell types between the OE and its glomerular target within the OB, which include both glial and neuronal cells. Neuronal cells within the OB with which the axons make synaptic contact include projection neurons and interneurons. However, using two targeted genetic deletions targeting each of these neuronal types, showed that OSN axons still coalesce in their normal glomerular location (Bulfone et al. 1998). This would indicate that signals from these neurons would not be a factor in axonal targeting and thus these cells do not contribute to convergence of the axons expressing one type of odorant receptor in the OB.

While the neurons in the OB appear to have seemingly no role in axon guidance, another cell type that interacts with the axons is believed to have a key role in axon extension: the olfactory ensheathing cell (OEC) (Ramón-Cueto and Avila 1998; Moreno-Flores et al. 2002; Barraud et al. 2013; Gómez et al. 2018). OECs are specialized glial cells that wrap around nerve fibers containing the axons of the OSNs both in the lamina propria and the ONL of the OB. The OEC is a unique glial cell type in that it has characteristics of both peripheral Schwann cells and central astrocytes while still having their own unique properties (Vincent, Taylor, et al. 2005). OECs are known to express and secrete multiple factors both *in vivo* and *in vitro* that have implications in axon guidance.

Two markers of OECs that can be seen in the lamina propria and distributed throughout the ONL are S100β (Franceschini and Barnett 1996; Au et al. 2002) and brain lipid binding protein (BLBP) (Rodriguez-Gil and Greer 2008), of which the former has been shown to be important for survival and axon extension (Van Eldik et al. 1991; Bhattacharyya et al. 1992; Li et al. 1998). Additionally, S100β has been shown to increase levels of amyloid precursor protein, which is another molecule that promotes outgrowth of neurites (Li et al. 1998; Moreno-Flores et al. 2003). An additional marker found expressed at low levels in OECs is glial fibrillary acidic protein (GFAP) (Barber and Lindsay 1982; Barber and Dahl 1987; Pixley 1992). GFAP is a component of intermediate filaments that is classically associated with astrocytes with which OECs do share some characteristics (Vincent, Taylor, et al. 2005). As it has been shown for astrocytes, increased levels of GFAP are detected in OECs after injury (Barber and Dahl 1987). While there are some OEC markers that seemingly are ubiquitously expressed, OECs are not a homozygous cell type in that they express distinct molecular markers that can divide the OECs into subpopulations both *in vitro* and *in vivo*.

In line with OECs sharing characteristics of both Schwann cells and astrocytes, OECs have been shown to take on the morphology of one of these two cell types when cultured *in vitro*. The astrocyte-like OECs show fibrous expression of GFAP as well as the embryonic, or polysialylated (PSA) form of the neural cell adhesion molecule (E-NCAM) *in vitro*. The Schwann-cell-like OECs show lower intensity of diffuse GFAP and no E-CAM *in vitro* but also show expression of the low-affinity p75 neurotrophin receptor

(p75NTR) (Pixley 1992; Franceschini and Barnett 1996). Additionally, molecularly

unique OECs can be seen *in vivo* depending on both the differentiation stage as well as

the location of the OEC (Vincent, West, et al. 2005). Distinct molecular markers of inner

and outer ONL have been described and could indicate different functions of OECs in

these two areas of the OB. Neuropeptide Y has been observed to be expressed by

OECs, with a greater intensity in the inner ONL (Ubink et al. 1994; Ubink and Hökfelt

2000). The expression of neuropeptide Y by OECs would further support a role in axon

guidance by OECs as neuropeptide Y has been shown to increase neurite growth as

well as to be an attractive cue (White and Mansfield 1996; White 1998; Hökfelt et al.

2008). OECs in the outer ONL are reported to express multiple markers that contrast it

from the inner ONL and have been shown to play a role in neurite outgrowth. These

include L1, E-NCAM, p75NTR, laminin, fibronectin, type IV collagen, and amyloid

precursor protein (Miragall et al. 1988; Miragall et al. 1989; Seki and Arai 1993;

Doucette 1996; Kafitz and Greer 1997; Kafitz and Greer 1999; Au and Roskams 2003;

Moreno-Flores et al. 2003; Gehler et al. 2004; Bonfanti 2006).

An additional glial cell type that the OSNs axons encounter are microglia.
Microglia are considered the resident immune cell of the brain and not well studied with

relation to axon growth. However, there has been recent work that has shown microglia

associated with developing axonal tracts (Pont-Lezica et al. 2014). Furthermore,

depletion of microglia using genetic approaches inhibited outgrowth of dopaminergic

axons in the developing mouse forebrain (Squarzoni et al. 2014). At this time, the

contribution of microglia to axon guidance in the olfactory system has not been described. However, microglia within the OB have been shown to play a role in synaptic pruning (Reshef et al. 2017; Wallace et al. 2020). Additionally, microglia have been reported to secrete neurotrophic factors and that they may play an indirect role in axon extension via modulation of PI3-K (Reemst et al. 2016).

Regeneration in the Olfactory System

Neurogenesis, or generation of newly born neurons from neural stem cells, is an area of continual research. It was long thought that the neurons that are present at birth are the only neurons present throughout life and no regeneration of neural cells occurred. However, it has since been shown that neurogenesis occurs in two different areas within the central nervous system: the subventricular zone and the dentate gyrus of the hippocampus (Kaplan and Hinds 1977; Luskin 1993; Kuhn et al. 1996; Doetsch et al. 1999). As previously mentioned, the subventricular zone is the source of neurogenesis that originates the rostral migratory stream, which leads to new interneurons within the OB (Whitman and Greer 2009). An additional source of neurogenesis within the olfactory system occurs within the OE and is critical for maintenance of the olfactory sense. Neurogenesis within the OE begins during embryonic development. After birth, neurogenesis within the OE progresses into a period of postnatal expansion before finally entering into a time of adult maintenance that continues throughout life (Murdoch and Roskams 2007).

Sensory Neuron Regeneration in the Olfactory Epithelium

OSNs have a high turnover resulting in an average lifespan of approximately 1 month. However, some OSNs can live from 3-12 months (Deckner et al. 1997). A contribution to the high turnover rate is OSNs at all stages of differentiation, including neural progenitor, immature, and mature OSN stages, undergo regular apoptosis in the mouse OE in a way that is believed to be a way to control the number of OSNs in the OE (Holcomb et al. 1995). Another contribution to the high turnover of OSNs is their exposure to the external environment in order to detect odorant molecules, which exposes the OSNs to potentially harmful elements (Ashwell 2012). The constant death of OSNs is counteracted by the continual neurogenesis that generates new OSNs. It is believed that the programmed cell death of OSNs and neurogenesis in the OE occurs throughout life. However, it has been shown that both apoptosis and proliferation decrease with age in the OE (Fung et al. 1997). In conflict with this report, an increase in apoptotic markers has been observed in the aged (24 month) rat OE (Robinson et al. 2002), which could play a role in age-related loss in olfaction. Additionally, it has been reported that the basal cells that are responsible for OE neurogenesis can become exhausted in both humans and in mice, which could further contribute to decreased neurogenesis and loss of the olfactory sense with age (Largent et al. 1993; Holbrook et al. 2005).

Horizontal vs Globose Basal Cells. A critical component of the regenerative capacity of the olfactory system are the basal cells in the OE that act as the progenitors for regeneration of the cells within the OE. In addition to regenerating OSNs, the basal cells can replenish non-neuronal cell types within the OE (Chen et al. 2004). As mentioned earlier, the basal cells can be divided into two categories: horizontal basal cells (HBCs) and the globose basal cells (GBCs). These two basal cell types are candidate stem cells and differ in that GBCs are actively proliferating while HBCs are largely quiescent (Mackay-Sim and Kittel 1991). The concurrence of proliferating and quiescent stem cells has been reported in other systems and is thought to play a key role in maintenance of these systems (Linheng and Clevers 2010).

Tissue stem cells have shared characteristics in that they are self-renewing, quiescent under normal conditions, and totipotent with respect to the tissue in which the stem cells are located (Pevny and Rao 2003; Schwob and Costanzo 2010). Within the OE, the HBCs seem to meet these requirements and thus could be considered the stem cells of the OE. HBCs appear to remain quiescent outside of an injury in which GBCs are destroyed and then HBCs can transiently proliferate to create new GBCs to regenerate the OE (Leung et al. 2007).

Lineage tracing studies have shown that GBCs are the direct progenitors of newly born neurons in the OE (Caggiano et al. 1994). In support of this, an area of the OE that is depleted of GBCs and suffers an injury will be reconstituted into respiratory epithelium, further demonstrating that GBCs are required for the regeneration of OSNs (Jang et al. 2003). Initially, GBCs were characterized by the absence of staining with markers that identified other cell types such as the HBC or OSN (Goldstein and Schwob 1996). Since that time, much has been learned about the GBC and is now believed to exist in multiple stages of differentiation and division state (Schwob and Costanzo 2010). One of these stages include a tentative stem cell state that appears to remain in a quiescent state, much like the HBCs (Jang et al. 2014). A second stage is the transient amplifying stage in which the GBCs divide to create a second GBC for self-renewal and then a second cell that will act as the immediate neuronal precursor (INP). Upon reaching the INP stage, the GBC is committed to the pathway of differentiation into an OSN (Huard et al. 1998).

Neuronal Differentiation and Axon Extension. The neural lineage of OSNs enters the INP phase after dividing from a transient amplifying GBC. The INP phase can be divided into two stages of which can be characterized by the expression of neuronal differentiation promoting transcription factors such as Ascl1 (Achaete-scute homolog 1 or Mash1) and neurgenin1 (Cau et al. 2002). Subsequent to the stages in which these transcription factors are expressed, the INP will undergo further differentiation into an immature and finally a mature OSN. Immature and mature OSN cell bodies can be differentiated based on their location within the OE as well as by distinct molecular markers. Immature OSN cell bodies are generally localized in the lower portion of the OE just above the basal cell layer as the cell bodies migrate apically during differentiation. The immature OSNs are characterized by the expression of growth-associated protein 43 (GAP43). As the OSNs further differentiate, their cell bodies further migrate more apically and will no longer express GAP43 but will now express olfactory marker protein (OMP), a classical marker of mature OSNs (Rodriguez-Gil et al. 2015; Liberia et al. 2019).

While markers such as GAP43 and OMP have long been used to differentiate immature and mature neurons within the OE, the process of axon extension after differentiation from neural progenitors proved more challenging. However, more recent work using genetic Cre-LoxP models have allowed for the timeline of axonal extension during OSN maturation to be defined (Rodriguez-Gil et al. 2015; Liberia et al. 2019). Additionally, this work on the process of OSN differentiation has shown that the onset of OR expression does not occur until the OSN axons reach the olfactory bulb. This would suggest that the OSN axon extension occurs in two phases, an OR-independent phase from the OE to the OB and an OR-dependent phase within the ONL of the OB to the glomerulus where the axon makes synaptic contact (Rodriguez-Gil et al. 2015). Furthermore, comparison of early post-natal and juvenile maturation of OSNs indicated that the progression of maturation displayed a similar timeline with only small delays in progression in juvenile subjects (Liberia et al. 2019).

Regeneration after an Olfactory Injury

In addition to OSN regeneration under normal conditions, the olfactory system also has the unique ability to recover and regenerate after an injury. The capacity of the OE to regenerate has been significantly studied using different models of an olfactory injury (Brann and Firestein 2014; Yu and Wu 2017). These injury models have been used to study physiological and behavioral responses within the olfactory system.

Olfactory Injury Models. A common experimental injury model is the removal of the OB, or olfactory bulbectomy (OBX). When an early postnatal OBX is performed, there is retrograde degradation of the OSNs in the OE. Interestingly, though the normal target tissue of the OSNs is absent, the OSNs extend new axons and form glomeruli within the cerebral hemisphere that has migrated anteriorly into the position normally occupied by the OB (Graziadei et al. 1978; Graziadei et al. 1979). However, other studies looking at cellular dynamics after OBX have shown there is an incomplete regeneration of the OE (Costanzo and Graziadei 1983) and a decrease in mature OSN markers (Carr et al. 1998; McIntyre et al. 2010). In spite of the abnormalities associated with regeneration after OBX, odorant detection and discrimination can occur after the OBX procedure (Slotnick et al. 2004). The OBX model, while valuable to study cell regeneration in the OE, may not be suitable for studying axon regeneration because the removal of the OB also removes the normal target of the OSNs and therefore the trophic support from the OB. Furthermore, the OBX model would be problematic in describing behavioral recovery of olfaction due to the lack of normal CNS structures and abnormal regeneration. Additionally, OBX has been extensively used as a model to study depression (Song and Leonard 2005; Morales-Medina et al. 2017), potentially complicating behavioral analysis in this model.

An additional model of olfactory injury is a manual transection of the OSN axons between the OB and the cribriform plate (Yee and Costanzo 1995; Yee and Costanzo 1998). Experiments using this axotomy model analyzed a wide range of post-injury time points. OSN axons have been reported to reach the OB by 35 days after axotomy in a hamster (Morrison and Costanzo 1995). In mice, recovery of connections assessed by olfactory marker protein (OMP)-LacZ expression is observed 42 days after lesion (Kobayashi and Costanzo 2009) and OMP-expressing axons begin to innervate the glomeruli as early as 25 days after surgery (Graziadei and Monti Graziadei 1980). Some of the differences observed in these results could be accounted for variations in the efficiency of the nerve transection. Olfactory behavior also appears to recover around 35 days after axotomy as this is when a response to odorant stimuli in hamsters can be observed (Costanzo 1985). Furthermore, a recovery of olfactory detection and discrimination has been shown at 40 days post axotomy (Yee and Costanzo 1995; Yee and Costanzo 1998) but the discriminatory tasks do require additional reinforcement in order for injured animals to relearn (Yee and Costanzo 1998).

Finally, the last model to study insult-related regeneration is chemical ablation via administration of olfactotoxic agents that produces seemingly selective damage to the olfactory epithelium. These treatments include multiple compounds and methods of delivery. The first example of these chemical ablation injuries is a nasal lavage administration of Zinc Sulfate ($ZnSO_4$) and Triton X-100 (Baker et al. 1983; Burd 1993). A second method of inducing a chemical ablation injury is via inhalation of methyl bromide (MeBr) gas (Schwob et al. 1995). Finally, the third approach to induction of a chemical ablation is via injection of chemicals such as dichlobenil or methimazole (Brandt et al. 1990; Genter et al. 1995). An advantage of chemical lesions is that the target, the olfactory bulb, remains intact, and only the cells in the olfactory epithelium are lost, followed by a rapid proliferation of the basally located stem cells (Hurtt et al. 1988; Genter et al. 1995; Williams et al. 2004). This makes the chemical ablation an ideal approach to carefully study events of axon regeneration and extension.

Cellular Dynamics after an Olfactory Injury. Studies of cellular dynamics in the olfactory epithelium after methimazole administration have shown onset of beta-tubulin III (a marker of immature OSNs) expression 4 days after lesion, and OMP at 7 days (Suzukawa et al. 2011). An additional study using the methimazole injury model showed small numbers of OMP-expressing cells 5 days after injury followed by gradual increases at 8, 15 and 22 days post-injury (Ogawa et al. 2014). Using a transgenic mouse model in which LacZ is co-expressed with the odorant receptor M72, it has been shown that there are only a few M72-expressing cells 10 days after methimazole injection. By 25 days post-injury, the number of M72-LacZ cells significantly increased in the epithelium and was recovered to control levels by 45 days (Blanco-Hernández et al. 2012).

Axonal Debris Removal

It is estimated that humans turnover as much as 300 billion cells every day (Arandjelovic and Ravichandran 2015). In the central nervous system, approximately 50% of the generated neurons experience apoptosis during development (Burek and Oppenheim 1996). With such a huge turnover of cells and creation of cellular debris after cell death, mechanisms for clearance of debris must exist for cellular systems to function properly. Furthermore, the suppression of the inflammatory response has been demonstrated to result from the clearance of apoptotic cells (Huynh et al. 2002). Clearance of neuronal debris is thought to be essential for suppressing inflammation, keeping a healthy neuronal environment, and helping in recovery of function (Neumann et al. 2009). Several neurodegenerative diseases, such as Alzheimer's and Parkinson's disease, present an inflammatory component that is characterized by the activation of microglia (Wang et al. 2015; Schain and Kreisl 2017).

Due to the constant turnover of OSNs, debris from the dying neurons in both the OE and the OB is constantly being generated. The process by which this debris is removed in the OE and OB is not well understood. The cells that act as the phagocytes in the olfactory system and the molecular pathways by which engulfment of this debris occurs has not been fully characterized. Additionally, it is not well known if this process differs after an injury when compared to normal regeneration.

Cellular Phagocytosis. In the primary olfactory pathway, the evidence for the cell type responsible for removing axonal debris is inconclusive. Despite some evidence of OECs removing debris, other evidence points to microglia and or macrophage infiltration being responsible for the clearance of neuronal debris. Cultured OECs have the ability to engulf apoptotic neurons (Su et al. 2013). Furthermore, during early development of a mouse, OECs have been shown to be responsible for clearing the degenerating axons (Nazareth et al. 2015). On the other hand, microglia have been reported in the olfactory nerve (Smithson and Kawaja 2010) and have also recently been associated with the refinement of connections in the olfactory bulb ((Lehmann et al., 2018) not peer reviewed). Furthermore, in a study using the Triton X-100 nasal lavage injury model followed by infection with *Staphylococcus aureus*, Iba1+ cell (common marker of macrophage and microglia) infiltration was observed in both the OE and the OB and seem to be in an activated phenotype based on coexpression of iNOS (Herbert et al. 2012). However, it is difficult to distinguish if the infiltration and activation of Iba1+ cells in this study is due to the olfactory lesion or the bacterial challenge. Yet, it is important to note that microglia/macrophages can respond to an insult of the olfactory system.

Molecular Mechanisms of Phagocytosis. The process of phagocytosis has three different types of signals that have been described: find-me, eat-me, and digest-me signals (Sierra et al. 2013). Find-me signals have been shown to involve the release of ATP and UTP from dead cells in order to recruit phagocytes to the site (Honda et al. 2001; Davalos et al. 2005; Koizumi et al. 2007). Eat-me signals can be subdivided into two processes: tethering and engulfing (Sierra et al. 2013). Tethering involves the recognition of apoptotic cells through signals such as exposed phosphatidylserine, which is normally only located in the inner leaflet of the phospholipid bilayer of the cell membrane but can be exposed after a cell undergoes apoptosis (Fadok et al. 1992; Fadok et al. 2000). The engulfment process is mediated but multiple different receptors

and pathways that signal the internalization of apoptotic cells and debris (Sierra et al.
2013; Brown and Neher 2014; Vilalta and Brown 2018). Finally, digest-me signals are
involved in the maturation of the phagosome and degradation of phagocytosed
materials (Desjardins et al. 1994; Garin et al. 2001).

Within the central nervous system, phagocytic signals have been studied largely
in microglia. Processes of microglia have been shown to be dynamic in their
movements in a way that is suggested to be a surveillance of their environment
(Nimmerjahn et al. 2005). This surveillance by microglia can lead to a response to areas
of injury or apoptotic cells via known find-me signals, such as ATP and fractalkine
(Davalos et al. 2005; Noda et al. 2011). Response to the fractalkine signal can lead to
the release of milk fat globule EGF factor 8, which can bind to phosphatidyl serine on
apoptotic cells and debris and act as an opsonin for the tethering of the microglia via
integrin RGD motifs (Leonardi-essmann et al. 2005; Fuller and Van Eldik 2008; Fricker
et al. 2012). Multiple pathways for engulfment in microglia have been identified. These
pathways include engulfment by a complement mechanism that has been shown to play
a role in retinal synaptic pruning *in vivo* and neurite engulfment *in vitro* (Linnartz et al.
2012; Schafer et al. 2012). Additionally, receptors in the TAM (Tyro3, Axl, MERTK)
family have been shown to be involved in microglial phagocytosis (Grommes et al.
2008; Fourgeaud et al. 2016). The digest-me signals in microglia are not well
understood, but have been observed in invertebrates that microglia require ATPase
activity in the phagolysosome in order for neuronal debris to be degraded (Peri and
Nüsslein-Volhard 2008).

Knowledge of phagocytic mechanisms within the olfactory system is limited. Most of our knowledge of axon removal in this sensory system after degeneration comes from invertebrates. *Drosophila melanogaster* OECs express components of the Draper engulfment signaling pathway and that these are essential for the clearance of sensory neuron axons that degenerated (Doherty et al. 2009; Doherty et al. 2014; Musashe et al. 2016). Homologs to the Draper pathway mediate, at least in part, phagocytosis in the mammalian peripheral nervous system where glial cell precursors clear debris from apoptotic cells via the receptors Jedi-1 (mammalian homolog of Draper) and MEGF-10 (Wu et al. 2009). Additional proteins required for the engulfment process have been described, such as the adaptor protein CED-6 in *Caenorhabditis elegans*, which recently was shown *in vitro* that its mammalian homologue, GULP, is also required for engulfment (Sullivan et al. 2014). However, neither these engulfment proteins that are seemingly expressed in OECs nor mechanisms underlying microglial phagocytosis have been studied in the mammalian olfactory system.

Discussion

The regenerative capability of the olfactory system is quite remarkable in that it able to maintain an extremely complicated neural network intact despite the high turnover rate of OSNs. Furthermore, the ability to recover from a devastating injury to the OE is unmatched in other areas of the nervous system. Despite learning a great deal about regeneration in the olfactory system, the underlying mechanisms are still not well understood. There are still many fundamental questions that need to be answered, which include the following: What is the mechanism underlying OR selection and how is only one allele of all possible ORs expressed? What are the mechanisms by which OSN axons extend from the OE to the OB and within the OB to target their specific synaptic target? What are the mechanisms by which OECs promote axon extension and how can this translate to therapeutic treatments in other systems? What are the cellular and molecular mechanisms by which axonal debris is removed in the olfactory system that seems to avoid a chronic inflammatory state? How is the synaptic system of the olfactory system restored functionally and anatomically after a complete ablation injury that destroys nearly all OSNs? Understanding these fundamental questions underlying the regenerative capability of the olfactory system will provide a wealth of knowledge into the ability of a neural system to regenerate and can provide powerful insight into mechanisms that could be applied to other neuronal systems.

CHAPTER 2. AXON EXTENSION AFTER METHIMAZOLE-INDUCED INJURY EMULATES NORMAL REGENERATION BUT DISPLAYS A DELAY IN SYNAPTOGENESIS

Introduction

Olfactory sensory neurons (OSNs) are the cells within the olfactory epithelium (OE) that are responsible for the detection of odorants. In order to detect odorant molecules, the OSNs must be partially exposed to the external environment in the nasal cavity, which can lead to the exposure of harmful elements and subsequent death of the OSN (Farbman 1990). This continuous loss of neurons is counterweighed by an ongoing neurogenesis. There is a population of stem cells at the base of the OE that can generate new cells throughout life to populate the OE, including OSNs (Brann and Firestein 2014). Therefore, the sense of smell is maintained despite the high death rate of OSNs due to the counteraction of cell loss via neurogenesis from the basal stem cells. In addition to the normal regeneration that takes place, the OE is capable of recovering from severe lesion injuries. After an injury, the basal stem cells can produce an OE that is both structurally (Schwob et al. 1995; Schwob et al. 1999) and functionally (Blanco-Hernández et al. 2012) similar to that of the pre-injury OE.

After a new neuron is born during either normal regeneration or after injury, it must extend an axon from the OE to the olfactory bulb (OB). In our previous work, we established the time course for axon extension during normal regeneration (Rodriguez-Gil et al. 2015) using a transgenic fate-mapping strategy. Nevertheless, data are scarce on neuron regeneration and axon extension when the olfactory system suffers a major injury. Our transgenic mouse model provides us an excellent tool to trace axon extension and therefore we decided to use it to understand injury-induced regeneration. Multiple injury models of the olfactory system have been employed and have shown varying results. Transection of the olfactory nerve has shown a large variance in

reinnervation of olfactory axons (Graziadei and Monti Graziadei 1980; Kobayashi and

Costanzo 2009), which could be attributed to the efficiency of the transection. OB

removal is another model used to study regeneration (Carr et al. 1998; McIntyre et al.

2010) but this removes the target of the newly growing axons and is therefore not suited

for the study of axon extension. The last approach we considered was chemical ablation

using olfactotoxic agents which can be delivered via nasal lavage, inhalation, or

injection (Hurtt et al. 1988; Genter et al. 1995; Williams et al. 2004). The chemical

ablation model is advantageous as the target, the OB, remains intact and only the cells

of the epithelium are removed apart from the basal stem cells, which begin to rapidly

proliferate after injury. We chose to use a chemical ablation model, specifically an

intraperitoneal injection of methimazole, in our study of axon extension after injury

because of its reproducibility. In this study, we use our developed Cre-LoxP model

(Rodriguez-Gil et al. 2015) to show the extension of axons from the OE to the OB, and

correlate this with behavior and synaptic connectivity after a methimazole-induced

injury.

Materials and Methods

Animals

    Ascl1tm1.1 (Cre/ERT2)Jejo/J mice (The Jackson Laboratory: stock# 012882)

were crossed with either B6.Cg-Gt(ROSA)26Sortm6(CAG-ZsGreen1)Hze/J reporter

mice (The Jackson Laboratory: stock# 007906) or B6.Cg-Gt(ROSA)26Sortm14(CAG-

tdTomato)Hze/J (The Jackson Laboratory: stock# 007914) *(Figure 2.1A)*. The use of

either reporter strain gave undistinguishable results under a double-blind analysis and the fluorescent reporter will subsequently be referred to as XFP as both reporter strains were used. Pups from these crosses were divided into two groups: some were treated at postnatal day (PND) 7 while others were allowed to age to 3-months-old. At their corresponding age, each animal received an intraperitoneal (IP) injection with either saline or methimazole (75 mg/kg, Sigma-Adrich: Cat# 46429). Pups that were heterozygous for both Ascl1 driven Cre-ERT2 and the fluorescent reporter (ZsGreen or TdTomato) were used for neuronal fate-mapping experiments. Littermates negative for the Ascl1 driven Cre-ERT2 were used for characterization of expression of other neuronal and synaptic markers post-injury. Double heterozygous pups received an IP injection 24 hours after saline or methimazole (at PND 8) with 4HO-tamoxifen (40 mg/kg, Sigma: Cat# H6278) or sunflower oil. In 24 hour increments post-injection, mice were anesthetized and transcardially perfused with PBS with 1 unit/mL of heparin, followed by 4% (wt/vol) paraformaldehyde (PFA) in PBS at 4°C. Tissue remained in 4% PFA overnight at 4°C for post-fixation and was then transferred to PBS. The tissue was then decalcified and then embedded in OCT compound and kept at −80 °C until use. Tissue collection times ranged from 2 to 40 days post saline/methimazole injection (1 to 10 days post 4HO-tamoxifen injection for axonal fate-mapping experiment; Figure 2.1B). All procedures were performed in accordance with the National Institute of Health Guide for the Care and Use of Laboratory Animals and were approved by the Institutional Animal Care and Usage Committee of East Tennessee State University.

Immunohistochemistry

Tissue sections were collected at 20 µm and stored at -20°C until use. On the day of use, the tissue sections were treated with blocking solution, 2% bovine serum albumin in PBS with 0.3% Triton X-100 (PBS-T) for 1-2 hours at room temperature. The slides were then incubated with primary antibodies diluted in the same blocking solution and left in a humidified chamber at 4°C for 1-2 nights. Primary antibodies and their dilutions were as follows: rabbit anti-ZsGreen (1:300, Clontech: Cat# 632474), goat anti-TdTomato (1:1000, Acris: Cat# AB8181-20), goat anti-OMP (1:2000, Wako: Cat# 544-10001), rabbit anti-TH (1:1000, Millipore: Cat# AB152), mouse anti-synaptophysin (1:1000, Chemicon Int: Cat# MAB329) mouse anti-VGlut1 (1/50, Synaptic Systems: Cat# 135 511), and rabbit anti-VGlut2 (1/200, Synaptic Systems: 135 403) . After incubation with primary antibodies, tissues were washed with PBS-T. Tissue was then incubated with secondary antibodies conjugated to Alexa Fluor and DAPI (1:1000) in a humidified chamber for 1-2 hours at room temperature. After this, tissue sections were washed with PBS-T and then PBS. Slides were covered with Fluoro-Gel mounting medium (EMS: Cat# 17985-11) and coverslipped. Images were acquired using an Olympus BX41 fluorescent microscope or a Leica SP6 confocal microscope.

Image Quantification

Measurement of olfactory marker protein (OMP) and XFP colocalization was done using the colocalization plugin for ImageJ software. The colocalized area within the olfactory epithelium was quantified and expressed as the area of colocalization divided by the total area of XFP staining within the OE. This procedure was performed on tissue collected on 8-10 and 21-days post 4HO-tamoxifen injection and 9-11 and 22 days after either methimazole or saline injection to compare the expression of OMP in newly born neurons after injury with normal regeneration.

Immunofluorescent staining of OB tissue collected 2-, 7-, 14- and 40-days post methimazole or age matched controls was performed for tyrosine hydroxylase (TH), synaptophysin, vesicular glutamate transporter (VGlut) 1, and VGlut2 as described above. Images were captured using a fluorescent microscope and analyzed using ImageJ software. In order to minimize differences in between experiments and samples, the level of fluorescence detected in the ONL was used to normalize background signal. After background fluorescence correction of each image, the mean fluorescence of TH, synaptophysin, VGlut1 and VGlut2 were detected within the glomeruli and quantified.

Hidden Cookie Test

3-month-old mice were used for assessment of their ability to find a hidden cookie after injury. Mice received either an intraperitoneal injection of methimazole (75 mg/kg) or saline and were tested at 3-, 14-, 21-, or 40-days post injection. Mice were fasted for 23 hours prior to testing. Mice were then placed in a 46 cm L X 23.5 cm W X 20 cm H cage with 5 cm of bedding and allowed to habituate for 10 minutes. After habituation, the mice were removed, and a cookie was hidden under the bedding. The mice were then placed in the center of the cage and the latency to find the cookie were recorded. The maximum time allowed to find the cookie was 15 minutes and a time of 900s was assigned when a mouse failed to find the cookie. Additionally, a double-blind analysis was performed to determine the latency to the target zone surrounding the cookie. The first occurrence of a mouse entering the area of the hidden cookie and remaining for a minimum of two seconds to rule out chance crossings was ruled as a positive entrance. Saline-injected controls run at 3-, 14-, 21-, and 40-days post injection were found to have no significant difference in latency to find the cookie and were grouped together in the final analysis.

Statistics

A Shapiro-Wilk normality test analysis of the behavioral and synaptic vesicle imaging numbers showed that the data was not normally distributed and therefore non-parametric statistical analyses were performed. Kruskal-Wallis one-way ANOVA was used to determine if there was a significant difference among multiple groups. For post-hoc analysis, Dunn's multiple-comparison test was used to compare MM-treated groups versus saline-treated controls. For OMP-XFP colocalization analysis, a two-way ANOVA was used to compare both injury versus normal regeneration and days post-4HOtamoxifen injection as independent variables. Post-hoc analysis using Student-Newman-Keuls test was performed to analyze difference between groups. A significant difference was indicated by a p-value of less than 0.05. Data is represented as mean ± SEM.

Results

Regeneration of the Olfactory Epithelium

To determine the best time for the 4HO-tamoxifen (4HO-Tx) injection, we administered methimazole at PND 7 and then injected 4HO-Tx at 24-, 48-, or 72-hours after methimazole administration and collected the tissue at PND 12. All three timepoints analyzed rendered high numbers of labeled OSNs that seemed undistinguishable from one another (data not shown). Given that we wanted to correlate axon extension and synapse formation with animal behavior, we selected the earliest timepoint (24 hours) for induction of the fluorescent marker after methimazole. We first set out to determine the temporal profile of injury-induced neuronal regeneration. In

order to do this, we looked at the labeling of OSNs cell bodies in the OE. At the first

timepoint of tissue collection, which was two days post methimazole (2 DPMM) or saline

injection (2 DPsal) and one day post 4HO-Tx injection (1 DPTX), we observed cells

labeled with XFP in both methimazole and saline treatments were located adjacent to

the lamina propria in the most basal portion of the epithelium (*Figure 2.1C-D*). This was

independent of the marked and expected difference in the thickness of the epithelium

between methimazole- and saline-injected mice. In addition, a portion of the shredded

epithelium could be seen on the apical surface of the newly growing epithelium in many

areas of the olfactory mucosa (*Figure 2.1C*). Additionally, as previously reported

(Rodriguez-Gil et al. 2015; Liberia et al. 2019), the labeled neurons largely contain

neither apical dendrites nor axons at this point. However, when we looked at labeled

cells at 2 DPTX, we saw a higher rate of extension of apical dendrites and axons

extending from the cell bodies and entering the lamina propria in both saline and

methimazole injected subjects (*Figure 2.1E-F*). It was noted in methimazole treated

subjects that the extension of the apical dendrites was sometimes horizontal through

the epithelium rather than vertically toward the apical surface of the epithelium as seen

in saline treated subjects. In subsequent days, the neuronal cell bodies migrated

apically toward the nasal cavity (*Figure 2.1 G-L*). Furthermore, methimazole treated

subjects showed an increase in thickness of their epithelia by 10 DPTX that approaches

that of the saline treated subjects (*Figure 2.1 K-L*).

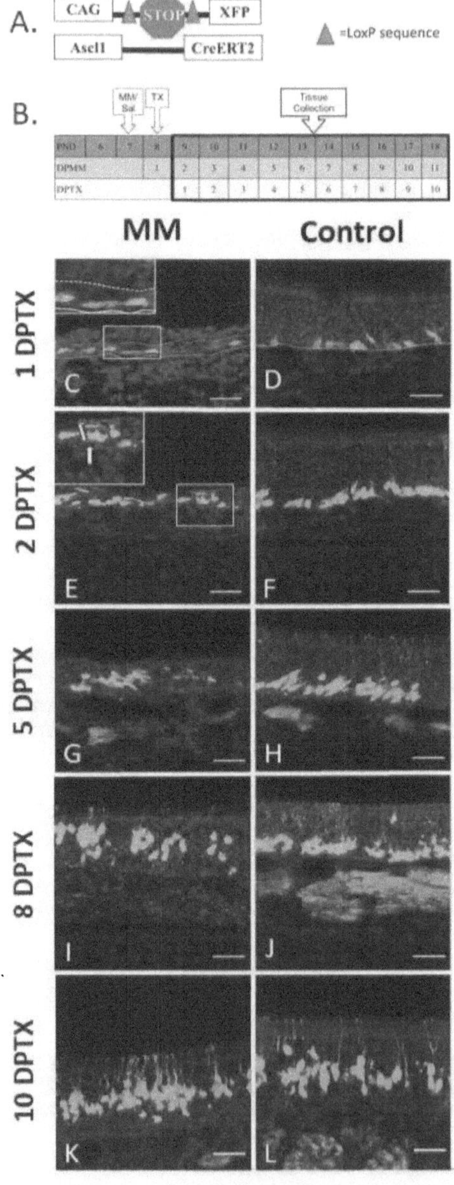
A.
CAG | STOP | XFP
AseI1 | CreERT2
= LoxP sequence
B.
MM/Sal. TX
Tissue Collection
PND | 6 | 7 | 8 | 9 | 10 | 11 | 12 | 13 | 14 | 15 | 16 | 17 | 18
DPMM | | 1 | 2 | 3 | 4 | 5 | 6 | 7 | 8 | 9 | 10 | 11
DPTX | | | 1 | 2 | 3 | 4 | 5 | 6 | 7 | 8 | 9 | 10
MM
Control
1 DPTX
C
D
2 DPTX
E
F
5 DPTX
G
H
8 DPTX
I
J
10 DPTX
K
L

Figure 2.1. Study design and immunofluorescent images showing the labeling of newly born neurons in the OE in control and methimazole (MM) treated subjects. *A*: Diagram of genetic design of CreERT2 driven by Asl1 promoter and either ZsGreen or TdTomato (represented as XFP) behind the strong CAG promoter and a LoxP flanked stop codon. *B*: Diagram showing strategy for injection of MM/Sal and Tx and subsequent tissue collection. *C-D*: At 1 DPTX, labeled neurons are located near the basement membrane of the OE indicated by the solid white line. A piece of shredding OE (indicated by the dashed line) is seen above the regenerating OE in the MM-injected (A, shown at higher magnification in the inset). *E-F*: At 2 DPTX, extension of axons and dendrites from the newly formed neurons can be seen. Higher magnification to show axons extending from cell bodies (*E inset*). *G-L*: From 5 to10 DPTX we see migration of cell bodies toward the apical surface of the OE. Additionally, the MM treated subjects show a recovery of the thickness of the epithelium that nears that of controls by 10 DPTX. (Scale bar: 30μm)

Extension of OSN Axons to the Olfactory Bulb

Next, we analyzed the extension of newly generated OSN axons to the olfactory bulb. Labeled axons in both methimazole and saline-treated subjects extended in a manner in which they reached the same landmarks at the same point post- 4HO-tamoxifen injection. Labeled axons were first seen crossing the cribriform plate into the OB at 3 DPTX (*Figure 2.2 A-B*). In the following days, the fluorescently labeled axons progressed deeper into the OB with the newly outgrowing axons located primarily in the outer olfactory nerve layer by 5 DPTX (*Figure 2.2 C-D*). By 8 DPTX, the axons had progressed from the outer olfactory nerve layer to the inner olfactory nerve layer and had begun to extend into the glomerular layer in the most ventral portion of the OB (*Figure 2.2 E-F*). At 10 DPTX, the glomeruli of the most ventral portion of the olfactory bulb were heavily labeled with fluorescently labeled axons (*Figure 2.2 G-H*). The time course of axon extension observed in these experiments is consistent what was previously reported in a non-injury model of axon extension (Rodriguez-Gil et al. 2015).

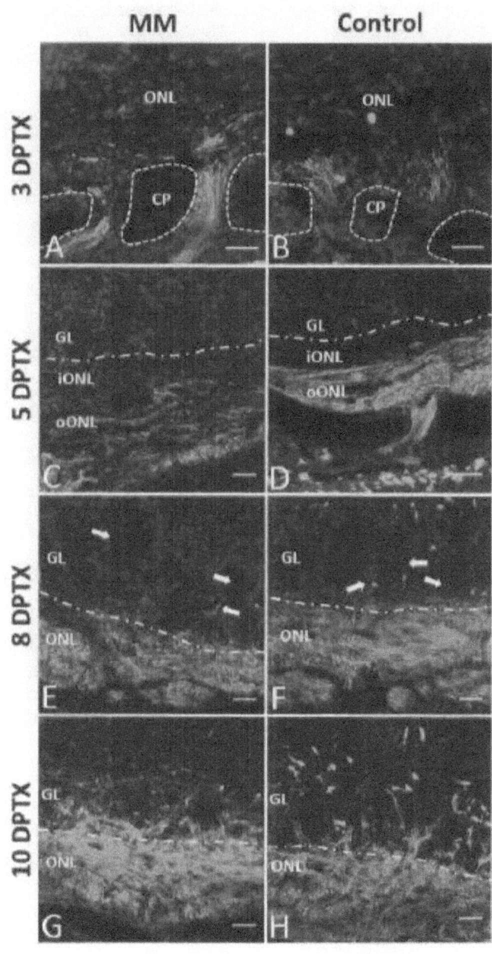

Figure 2.2. Immunofluorescent tracing of labeled OSN axons in control and methimazole (MM) treated subjects. *A-B*: At 3 DPTX, labeled axons are first seen crossing the cribriform plate and entering the ONL of the OB. *C-D*: At 5 DPTX, axons are primarily in the outer olfactory nerve layer (oONL). *E-F*: At 8 DPTX, axons

encompass the entire ONL and have begun to enter the glomerular layer (GL) (white arrows). *G-H*: At 10 DPTX, the axons are heavily present in the glomeruli. (Scale bar: 30µm) OE: olfactory epithelium, CP: cribriform plate, ONL: olfactory nerve layer, oONL- outer ONL, iONL- inner ONL, GL- glomerular layer

OMP Expression in Newly Born Neurons of the OE

To assess maturation of the newly born OSNs after injury, the co-expression of OMP with XFP in OSNs labeled after injury was analyzed and compared to co-expression of OMP and XFP in OSNs labeled during normal regeneration (*Figure 2.3*). Control OSNs showed colocalization of OMP and XFP at 8-, 9-, and 10 DPTX but did not show an increase in their levels of colocalization from 8-10 DPTX. In contrast with control OSNs, the colocalization of XFP and OMP in MM treated OSNs showed a trend suggestive of an increase in colocalization between 8-10 DPTX (*Figure 2.3B*), but statistical analysis revealed there was no statistical difference between these timepoints. By 21 DPTX, a larger co-expression of OMP and XFP was seen in both control and MM-treated OSNs and colocalization at this timepoint was significantly higher than that seen at 8-, 9-, and 10DPTX (*Figure 2.3B*). Analysis of interaction between the independent factors of injury versus normal regeneration and DPTX show no interaction between these two factors, indicating that MM treatment had no effect on maturation of newly born OSNs as indicated by co-expression of OMP in the newly born OSNs.

A.

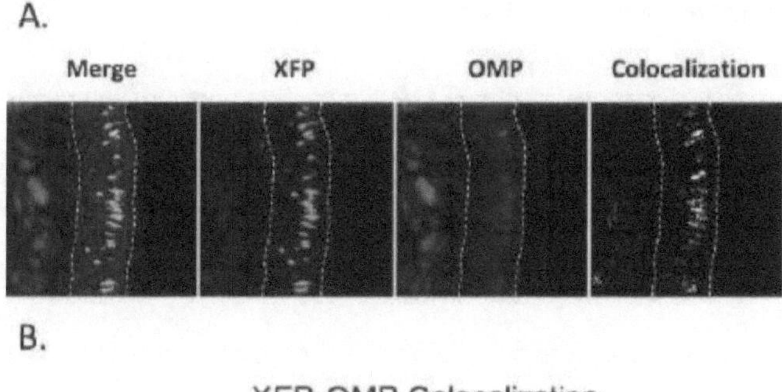

B.

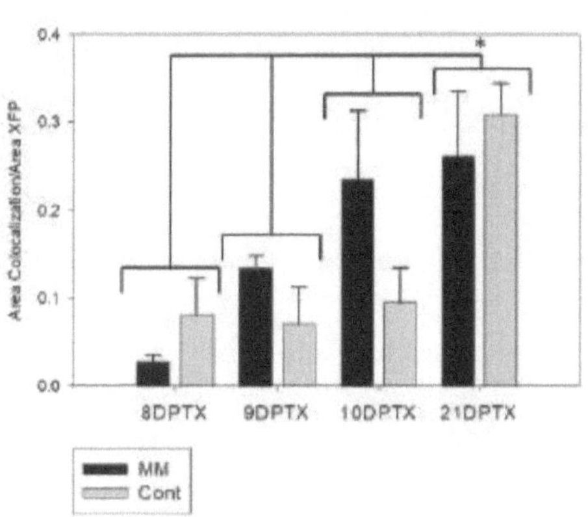

Figure 2.3. Colocalization of XFP and OMP in the OE. *A.* Representative image showing immunofluorescent staining of the OE with XFP and OMP and the colocalization of the two markers. B. Quantification of the colocalization analysis comparing control and methimazole-treated groups. Two-way ANOVA shows that 21DPTX is significantly different from 8-, 9-, and 10DPTX but there is no difference between MM and control groups. N=3 for each group, * p≤0.05.

To investigate changes in synaptic activity and synaptic density after injury, levels of TH and synaptophysin were analyzed within tissue sections of the olfactory bulb. TH was used as an indicator of synaptic activity as TH expression in periglomerular cells is correlated with activity within the glomeruli (Sakai and Nagatsu 1993). Synaptophysin is a protein expressed in synaptic vesicles of the glomerular layer (GL) and the external plexiform layer (Kasowski et al. 1999) where it is located in synaptic terminals of OSNs, periglomerular cell, and projection neurons (mitral and tufted cells), and it was used to assess overall synaptic density in the GL (*Figure 2.4A-C*). Analysis of the intensity of synaptophysin and TH staining within the GL indicates no statistically significant changes at any timepoint post-injury when compared to controls.

To assess synaptic markers that were specific to dendritic processes and the axon terminals of the OSN, VGlut1 and VGlut2 levels within the GL were assessed (*Figure 2.4D-F*). VGlut1, which specifically labels the dendritic processes of the mitral/tufted cells, shows a significant decrease at 2-, 7-, and 14-days post-MM treatment (*Figure 2.4E*). Additionally, VGlut2, which specifically labels axon terminals of OSNs, shows a significant decrease at 2- and 14-days post-MM treatment (*Figure 2.4F*). There is a return to control levels at 7 DPMM, which could coincide with the initial arrival of new axons and could reflect both the establishment of new synaptic connections along with remaining axonal debris leading to a seeming increase in VGlut2 signal within the GL. The following decrease at 14-days post-MM could be attributed to removal of dead terminals or to a refinement of new connections, as it has been shown that after injury there is a high level of mistargeting by OSNs (Holbrook et al. 2014). The levels of both VGlut1 and VGlut2 return to that of age-matched controls by 40 DPMM,

which indicates a delay in the recovery of synapses within the GL when compared to

the initial arrival of OSN axons shown in our fate-tracing experiments using XFP labeled

axons.

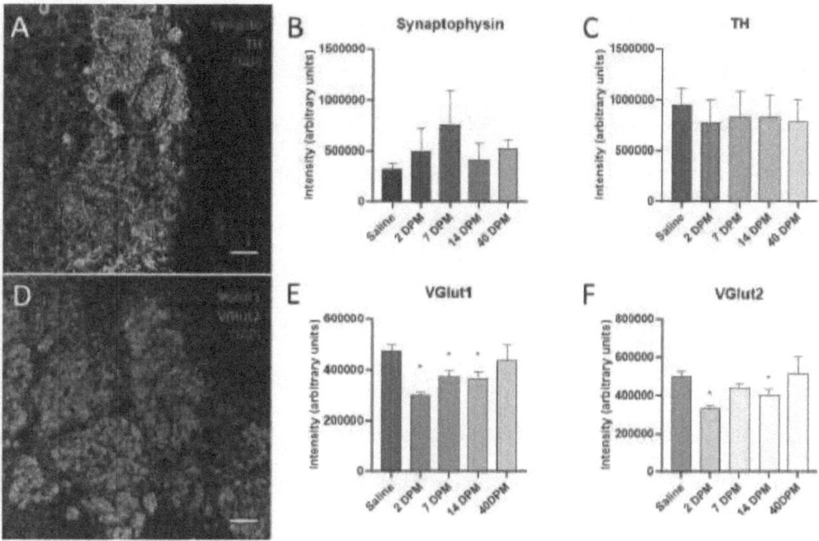

Figure 2.4. Synaptic markers in the GL. *A.* Representative image of synaptophysin (red)

and TH (green) staining of the GL. *B.* Synaptophysin quantification. *C.* TH

quantification. *D.* Representative image of VGlut1 (red) and VGlut2 (green) staining of

the GL (Note there is little yellow indicating that these two markers are located in

subcellular compartments of two different cell types: dendritic processes of projection

neurons and axons of OSNs respectively). *E.* VGlut1 quantification. *F.* VGlut2

quantification. * $p \leq 0.05$ vs. control

Hidden Cookie Test

Finally, to correlate these histological hallmarks with functional recovery of the olfactory system after an injury, the ability of a mouse to find a hidden cookie was assessed. Due to preliminary observations of behavior during this test, we also performed a double-blind analysis of the latency of a mouse to spend a pre-defined time in the target zone of the hidden cookie. At 3 DPMM, assessment of latency to find the hidden cookie showed a significant deficit ($p<0.05$) when compared to saline treated controls (*Figure 2.5A*). At 14 DPMM, a reduction in the time to find the hidden cookie was observed when compared to 3 DPMM, but this time was still significantly different ($p<0.05$) than the saline treated controls. Subsequent behavioral testing at 21 and 40 DPMM revealed a reversal of the trend toward recovery to control levels and showed an increase in latency to find the cookie (*Figure 2.5A*). A double-blinded analysis then analyzed latency to the first visit to the cookie zone and found that it mimics the trend on latency to find the cookie of 3- and 14 DPMM in which a significant deficit is seen at 3 DPMM and there is a trend towards recovery by 14 DPMM which is still significantly different ($p<0.05$) from control levels (*Figure 2.5B*). However, this measure shows a continuation of the trend toward recovery to control levels at 21- and 40 DPMM and the latency at 40 DPMM to the cookie zone is no longer significantly different from control levels. The discrepancy between both assessments at 21- and 40 DPMM could be due to behavioral changes that are not olfactory in nature but could be associated with long term side effects of the MM treatment affecting motivation perhaps through interaction with the thyroid (Cai et al. 2018; Smith and Northcutt 2018).

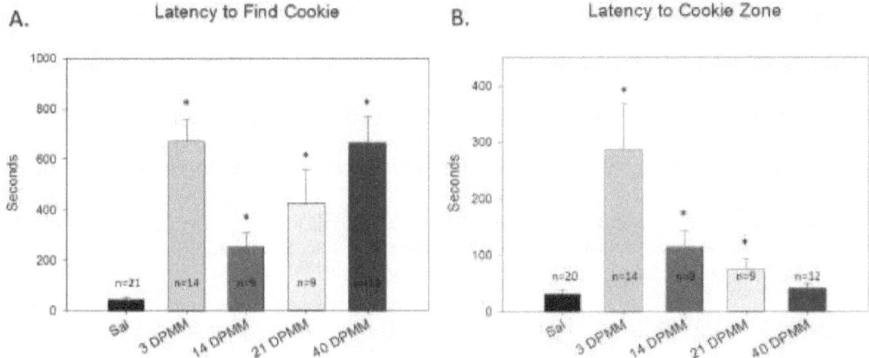

Figure 2.5. Hidden Cookie Test. *A.* Latency to find cookie. *B.* Latency to the cookie zone. * P≤0.05 vs control.

Discussion

The results of this study suggest that despite the shedding of virtually the entire OE after a MM-induced injury, the timeline for axon extension from the OE to the OB is unchanged when compared to normal regeneration. This would suggest that mechanisms behind axon guidance remain intact even after a chemical ablation insult. Establishing a timeline for extension of OSN axons is critical for the study of mechanisms underlying axon extension and targeting. The spatiotemporal precision provided by this Cre-LoxP fate-tracing model used in this study and others (Rodriguez-Gil et al. 2015; Liberia et al. 2019) have established a distinct time course of events and landmarks for axon extension that can be used in conjunction with mechanistic studies of OSNs at given timepoints after division from Ascl1+ neural precursors.

While this study found that axons do reach the GL at around 8-10 DPTX (or 9-11DPMM), our results also indicate a delay in the recovery of synaptic connections within the OB. This is further demonstrated by the delay in recovery of olfactory behavior after injury. It has previously been reported that axonal targeting is less accurate post-injury (Holbrook et al. 2014), which could suggest that delayed anatomical and functional recovery is due to refinement that must occur in the synaptic network of the OB due to inaccurate axonal targeting. Furthermore, the early (but not significant) rise in colocalization of the newly generated OSNs with OMP after MM treatment could be indicative of the refinement process that must occur after injury as OMP has recently been shown to play a role in the refinement of the OB circuitry (Albeanu et al. 2018).

The results of the hidden cookie test in MM-treated mice indicates that mice do display abnormal behavior, even at later timepoints post-injury. However, the latency to the target zone analysis suggests that this altered behavior is not olfactory in nature. Furthermore, the inhibition in their ability to find the hidden cookie did not appear to involve altered hunger in the mice as they had been fasted for 23 hours prior to testing and after testing would eat the cookie when it was removed from the bedding for them. All of these observations would suggest that a long-term off-target effect from the MM treatment is present and, in some way, affects the motivation of the mice to dig for the cookie during testing. MM is used clinically in the treatment of hyperthyroidism and alterations in motivational behavior have been reported due to alterations in the thyroid (Cai et al. 2018; Smith and Northcutt 2018). However, the mechanisms behind this altered behavior is beyond the scope of the current study but would suggest the MM olfactory injury model is more suited for histo-anatomical studies of olfactory injury rather than behavioral analysis of olfactory behavior post-injury.

In summary, the key finding of this study is that despite the arrival of axons to the GL at around 10 days post-injury, there is a delay of up to 40 days for the synaptic network refinement. The factors determining the inaccurate targeting of axons and their subsequent pruning are unknown at this time. Future work can establish the cues that lead to off-target glomerular targeting as well as the cellular and molecular mechanisms underlying the pruning and refinement of the OB synaptic network.

References

Albeanu DF, Provost AC, Agarwal P, Soucy ER, Zak JD, Murthy VN. 2018. Olfactory marker protein (OMP) regulates formation and refinement of the olfactory glomerular map. Nat Commun. 9(1):1–12. doi:10.1038/s41467-018-07544-9.

Blanco-Hernández E, Valle-Leija P, Zomosa-Signoret V, Drucker-Colín R, Vidaltamayo R. 2012. Odor Memory Stability after Reinnervation of the Olfactory Bulb. PLoS One. 7(10). doi:10.1371/journal.pone.0046338.

Brann JH, Firestein SJ. 2014. A lifetime of neurogenesis in the olfactory system. Front Neurosci. 8:182. doi:10.3389/fnins.2014.00182. [accessed 2019 Sep 16]. http://journal.frontiersin.org/article/10.3389/fnins.2014.00182/abstract.

Cai YJ, Wang F, Chen ZX, Li L, Fan H, Wu ZB, Ge JF, Hu W, Wang QN, Zhu DF. 2018. Hashimoto's thyroiditis induces neuroinflammation and emotional alterations in euthyroid mice. J Neuroinflammation. 15(299). doi:10.1186/s12974-018-1341-z.

Carr VM, Walters E, Margolis FL, Farbman AI. 1998. An enhanced olfactory marker protein immunoreactivity in individual olfactory receptor neurons following olfactory bulbectomy may be related to increased neurogenesis. J Neurobiol. 34(4):377–90. [accessed 2019 Sep 16]. http://www.ncbi.nlm.nih.gov/pubmed/9514526.

Farbman AI. 1990. Olfactory neurogenesis: genetic or environmental controls? Trends Neurosci. 13(9):362–365. doi:10.1016/0166-2236(90)90017-5. [accessed 2019 Sep 16]. https://www.sciencedirect.com/science/article/abs/pii/0166223690900175?via%3Dihub.

Genter MB, Deamer NJ, Blake BL, Wesley DS, Levi PE. 1995. Olfactory Toxicity of Methimazole: Dose-Response and Structure-Activity Studies and Characterization of Flavin-Containing Monooxygenase Activity in the Long-Evans Rat Olfactory Mucosa. Toxicol Pathol. 23(4):477–486.

doi:10.1177/019262339502300404. [accessed 2019 Sep 16].
http://journals.sagepub.com/doi/10.1177/019262339502300404.

Graziadei PPC, Monti Graziadei GA. 1980. Neurogenesis and neuron regeneration in
the olfactory system of mammals. III. Deafferentation and reinnervation of the
olfactory bulb following section of thefila olfactoria in rat. J Neurocytol. 9(2):145–
162. doi:10.1007/BF01205155. [accessed 2019 Sep 16].
http://link.springer.com/10.1007/BF01205155.

Holbrook EH, Iwema CL, Peluso CE, Schwob JE. 2014. The regeneration of P2
olfactory sensory neurons is selectively impaired following methyl bromide lesion.
Chem Senses. 39(7):601–616. doi:10.1093/chemse/bju033.

Hurtt ME, Thomas DA, Working PK, Monticello TM, Morgan KT. 1988. Degeneration
and regeneration of the olfactory epithelium following inhalation exposure to
methyl bromide: Pathology, cell kinetics, and olfactory function. Toxicol Appl
Pharmacol. 94(2):311–328. doi:10.1016/0041-008X(88)90273-6. [accessed 2019
Sep 16].
https://www.sciencedirect.com/science/article/pii/0041008X88902736?via%3Dihu
b.

Kasowski HJ, Kim H, Greer CA. 1999. Compartmental organization of the olfactory bulb
glomerulus. J Comp Neurol. 407(2):261–274. doi:10.1002/(SICI)1096-
9861(19990503)407:2<261::AID-CNE7>3.0.CO;2-G.

Kobayashi M, Costanzo RM. 2009. Olfactory Nerve Recovery Following Mild and
Severe Injury and the Efficacy of Dexamethasone Treatment. Chem Senses.
34(7):573–580. doi:10.1093/chemse/bjp038. [accessed 2019 Sep 16].
https://academic.oup.com/chemse/article-lookup/doi/10.1093/chemse/bjp038.

Liberia T, Martin-Lopez E, Meller SJ, Greer CA. 2019. Sequential maturation of olfactory
sensory neurons in the mature olfactory epithelium. eNeuro. 6(5):ENEURO.0266-
19.2019. doi:10.1523/eneuro.0266-19.2019.

McIntyre JC, Titlow WB, McClintock TS. 2010. Axon growth and guidance genes
identify nascent, immature, and mature olfactory sensory neurons. J Neurosci
Res. 88(15):3243–3256. doi:10.1002/jnr.22497. [accessed 2019 Sep 16].
http://doi.wiley.com/10.1002/jnr.22497.

Rodriguez-Gil DJ, Bartel DL, Jaspers AW, Mobley AS, Imamura F, Greer C a. 2015.
Odorant receptors regulate the final glomerular coalescence of olfactory sensory
neuron axons. Proc Natl Acad Sci.:201417955. doi:10.1073/pnas.1417955112.

Sakai M, Nagatsu I. 1993. Alteration of carnosine expression in olfactory system of
mouse after unilateral naris closure and partial bulbectomy. J Neurosci Res.
34(6):648–653. doi:10.1002/jnr.490340608. [accessed 2019 Sep 16].
http://doi.wiley.com/10.1002/jnr.490340608.

Schwob JE, Youngentob SL, Mezza RC. 1995. Reconstitution of the rat olfactory
epithelium after methyl bromide-induced lesion. J Comp Neurol. 359(1):15–37.
doi:10.1002/cne.903590103. [accessed 2019 Sep 16].
http://doi.wiley.com/10.1002/cne.903590103.

Schwob JE, Youngentob SL, Ring G, Iwema CL, Mezza RC. 1999. Reinnervation of the
rat olfactory bulb after methyl bromide-induced lesion: Timing and extent of
reinnervation. J Comp Neurol. 412(3):439–457. doi:10.1002/(SICI)1096-
9861(19990927)412:3<439::AID-CNE5>3.0.CO;2-H. [accessed 2019 Sep 16].
http://doi.wiley.com/10.1002/%28SICI%291096-
9861%2819990927%29412%3A3%3C439%3A%3AAID-
CNE5%3E3.0.CO%3B2-H.

Smith SG, Northcutt K V. 2018. Perinatal hypothyroidism increases play behaviors in
juvenile rats. Horm Behav. 98:1–7. doi:10.1016/j.yhbeh.2017.11.012.

Williams SK, Gilbey T, Barnett SC. 2004. Immunohistochemical Studies of the Cellular
Changes in the Peripheral Olfactory System After Zinc Sulfate Nasal Irrigation.
Neurochem Res. 29(5):891–901. doi:10.1023/B:NERE.0000021234.46315.34.
[accessed 2019 Sep 16].
http://link.springer.com/10.1023/B:NERE.0000021234.46315.34.

CHAPTER 3. ROLE OF MICROGLIA IN REMOVAL OF AXONAL DEBRIS IN THE

OLFACTORY BULB AFTER A METHIMAZOLE-INDUCED INJURY

Introduction

Olfactory sensory neurons (OSNs) are constantly dying. Despite the constant turnover of OSNs, the olfactory system has specific mechanisms by which the neural network responsible for the perception of smell can be regenerated and maintained. The regenerative capacity of the olfactory system is further enhanced by the ability to recover from an injury in which part or seemingly the entire olfactory epithelium (OE) is damaged (Schwob 2002).

Injuries to the olfactory system can be modeled in mice using chemical ablation of the OE via administration of olfactotoxic agents. These ablation type of injuries lead to the shedding of the OE, including the immature and mature OSNs, supporting cells, and some of the basal cells (Schwob and Costanzo 2010). It is estimated that the mouse OE contains around 10 million OSNs (Kawagishi et al. 2014). This number of neurons that can be lost in an ablation injury is huge when compared to the estimated 13 million neurons in the mouse cerebral cortex (Herculano-Houzel et al. 2006). Yet, despite this massive neural injury, the OE and synaptic network within the olfactory bulb (OB) is able to regenerate both functionally and anatomically to a state which is seemingly identical to the preinjury system (cf. Chapter 1) (Blanco-Hernández et al. 2012).

Chemical ablation type injuries lead to the shedding of the OE, which removes the apical dendrites and cell bodies of the OSNs. However, the axons in the lamina propria and OB remain after such an injury. The cellular phagocyte behind the removal of axonal debris left from a chemical ablation olfactory injury in a mammalian model has yet to be identified. It has been shown that olfactory ensheathing cells (OECs) phagocytose debris during early development (Nazareth et al. 2015). Furthermore, OECs have been shown *in vitro* to phagocytose neuronal debris (Su et al. 2013). However, an additional potential phagocyte that plays a role in phagocytosis in other systems, microglia, have been shown to be present in the olfactory nerve as well (Smithson and Kawaja 2010). It is the goal of this study to identify the role of microglia in the removal of axonal debris after a methimazole (MM) induced injury within the OB.

Materials and Methods

Animals

12-week-old C57BL/6 mice were ordered from Jackson laboratory and were treated with either saline or MM (75 mg/kg, Sigma-Adrich: Cat# 46429) and were collected at 3-, 14-, 21-, and 40-days post injection. Additionally, our Cre-LoxP model as previously described (Chapter 2 methods) was utilized. In brief, Ascl1^{CreERT2} mice (The Jackson Laboratory: stock# 012882) were crossed with either R26R^{ZsGreen} (The Jackson Laboratory: stock# 007906) or R26R^{TdTomato} reporter mice (The Jackson Laboratory: stock# 007914). Double-heterozygous pups from these crosses were used at early postnatal timepoints for axonal fate-tracing while Ascl1 wildtype pups were aged to 3 months and used for OB protein and RNA collection. Double-heterozygous mice were treated with 4HO-tamoxifen (40 mg/kg, Sigma: Cat# H6278) at post-natal day (PND) 7. After allowing axons to grow for 10 days to reach the glomerular layer of the OB (Chapter 2), the mice were injected with either Saline or MM at PND17. Mice were collected at 3-,7-, 14-, 21-, and 40-days post saline/MM injection. At time of collection, mice were euthanized and either and had their OB's removed and flash frozen using dry

58

ice or were transcardially perfused with 4% paraformaldehyde (PFA) as previously described (Chapter 2 methods). Perfused tissue was left in 4% PFA overnight at 4°C and subsequently decalcified and frozen in OCT. All tissue samples were left at -80°C until use. All procedures were performed in accordance with the National Institute of Health Guide for the Care and Use of Laboratory Animals and were approved by the Institutional Animal Care and Usage Committee of East Tennessee State University.

Immunohistochemistry

Immunohistochemistry was performed as described previously (Chapter 2). Briefly, 20 µm sections were blocked using 2% bovine serum albumin for 1-2 hours at room temperature. Primary antibodies were then added and left overnight at 4°C. Slides were then washed and either Alexa Flour or horse radish peroxidase (HRP) conjugated secondary antibodies and DAPI (1/1000) were added for 1-2 hours at room temperature. Stainings using HRP conjugated secondary antibodies were visualized using Alexa Fluor 488 tyramide reagent (Invitrogen: Cat#B40953). After subsequent washes, slides were covered with mounting medium and coverslipped. Primary antibodies and their dilutions were as follows: goat anti-TdTomato (1:1000, Acris: Cat# AB8181-20), rabbit ant-IBA1 (1/1000, Fujifilm: Cat# 019-19741), rat anti-CD11b (1/50, BD Pharmingen: Cat# 550282). Images were acquired using a Leica SP6 confocal microscope.

Western Blot

For Western blot analysis, OB samples were placed in RIPA lysis buffer (Millipore: Cat# 20-188) with protease and phosphatase inhibitor cocktail (Thermo Scientific: Cat# 78443) and homogenized using a handheld homogenizer. Samples were then left on ice for 30 minutes and then centrifuged at 12000 x g for 15 minutes at 4°C. Supernatant was then collected and protein quantification was carried out using a BCA assay (Thermo Scientific, Cat# 23227). Protein samples were diluted in Laemmli buffer plus 2-mercaptoethanol and denatured at 95°C for 5 minutes then loaded into a 15% agarose gel and run through SDS-PAGE. Protein was then transferred to a PVDF membrane and blocked with 5% milk in TBS-T. Membranes were then probed with a rabbit anti-IBA1 primary antibody (1/1000, Fujifilm: Cat# 019-19741) diluted in the blocking buffer and left on a rotator overnight at 4°C. After washes, an HRP conjugated anti-rabbit antibody was added for 1 hour at room temperature followed by washes. The PVDF membranes were then treated chemiluminescent HRP substrate (Millipore: Cat# P90719) and then imaged using a LI-COR Odyssey® imaging system. For a loading control, membranes were stripped and treated with rabbit anti-actin (1/1000, Cell Signaling: Cat# 13E5). Band density was quantified and used to compare protein expression between MM-treated OBs and saline-treated OBs.

RT-QPCR

RNA was isolated from frozen OB samples using TRIzol LS reagent (Ambion: Cat# 10296010) by following the manufacturer's instructions. RNA concentrations were determined using a Biotek Epoch 2 microplate reader and Biotek Take3 microvolume plate. RNA was then used in a reverse transcription reaction to create 10 ng/µL cDNA. Between 7.5 and 20 ng of cDNA were used in each PCR reaction. RT-qPCR reactions were performed using SybrGreen reaction mix (Thermo Scientific: Cat#AB-1323/A) and an Agilent Technologies Stratagene Mx3000P. PCR reactions took place using the following protocol: 10 minutes at 95°C; 40 cyclical repeats of 15 seconds at 95°C, 30 seconds at 62°C, and 30 seconds at 72°C; followed by a disassociation curve protocol to ensure the creation of only one PCR product.

Primers were designed such that the PCR product would span an intron when possible in order to amplify mRNA transcripts more specifically and lessen the chance of genomic DNA contamination being amplified. The following primers were used: Megf10 forward 5'-TGCAGGAGTCGTATCCACAT-3'; Megf10 reverse 5'-TGTCCGGTAGCTGATTCTGT-3'; Jedi1 forward 5'-CTCTCCCGTGACCCATAACT-3'; Jedi1 reverse 5'-GAGAGACATCTGGCATGACG-3'; Gulp forward 5'-GATCTGCTGAGACCAAACGG-3'; Gulp reverse 5'-CAAATTAAGCACGTCAGGATCAC-3'; CD11b forward 5'-AGGCTCTCAGAGAATGTCCT-3'; CD11b reverse 5'-CGTCCGAGTACTGCATCAAA-3'; Tlr2 forward 5'-TGTTTCTGAGTGTAGGGGCT-3'; Tlr2 reverse 5'-AAGAGCTCGTAGCATCCTCT-3'; Tlr4 forward 5'-AATCCCTGCATAGAGGTAGTTCC-3'; Tlr4 reverse 5'-TCTGGATAGGGTTTCCTGTCA-3'; C1qa forward 5'-GGAGCATCCAGTTTGATCGG-3'; C1qa reverse 5'-AAACCTCGGATACCAGTCCG-3'; Complement3 forward 5'-GTAGTGATTGAGGATGGTGTGG-3'; Complement3 reverse 5'-GATGACAGTGACGGAGACATAC-3'. These genes were normalized to the expression levels of 3 housekeeping genes: β-actin, β2 microglobulin, and TATA box binding protein. The primers used for the housekeeping genes are as follows: β-actin forward 5'-CAACGAGCGGTTCCGATG-3'; β-actin reverse 5'-GCCACAGGATTCCATACCCA-3'; β2 microglobulin forward 5'-TGTCTCACTGACCGGCCTGTATGCT-3'; β2 microglobulin reverse 5'-ATTCTCCGGTGGGTGGCGTGAGTAT-3'; TATA forward 5'-

TGGACCAGAACAACAGCCTTCCAC-3'; TATA reverse 5'-

CCTGTGCCGTAAGGCATCATTGGA-3'.

Statistics

Data across multiple groups were compared using one-way ANOVA. Data from qPCR experiments lacked a normal distribution and was analyzed using a Kruskal-Wallis one-way ANOVA. Dunn's multiple-comparison test was performed to compare MM-treated groups versus saline-treated controls. A p-value of less than 0.05 was used to identify a significant difference. Graphical representation of data is displayed as mean ± SEM.

Results

Influx of microglia after MM injury

Our goal in this study was to determine if microglia play a role in the removal of axonal debris after a MM-induced injury. We performed Western blot studies on OB tissue to survey levels of a common microglial marker, IBA1, after injury. Protein levels do appear to increase at 3, 14, and 21 DPMM in the OB and return to levels seen in saline treated animals by 40 DPMM. However, quantification of these results reveals these changes are not statistically significant (*Figure 3.1*).

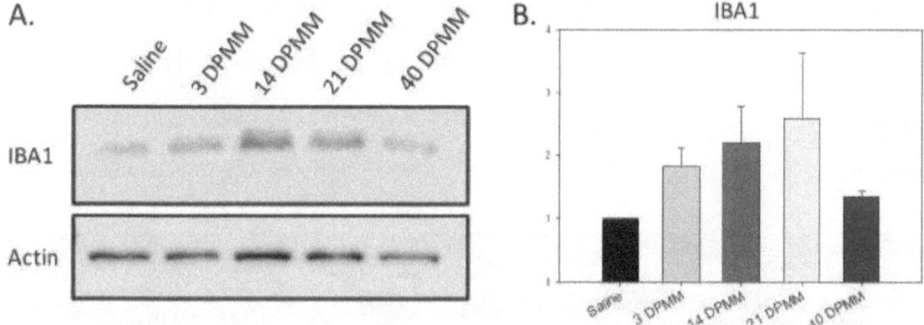

Figure 3.1. Western blot of IBA1 in OB tissue post-injury. *A.* Representative IBA1 blot showing increase in IBA1 in OB tissue post injury. *B.* Quantification of these blots reveal that the change is not significantly different from controls (p=0.097). N=4

While these results did not produce statistically significant results, it was determined that an increase in IBA1 was possibly underestimated in our area of interest because we were using homogenates from the entire OB. The OSN axons are located on the two outermost layers of the OB, the olfactory nerve layer (ONL) and the glomerular layer (GL). Therefore, it is possible that an increase within these two layers could be missed by using the entire OB for Western blotting. With this in mind, we performed immunofluorescent staining on OB tissue post-injury to achieve a higher resolution of the changes of IBA1+ cells on the individual layers of the OB. Immunofluorescent staining revealed that IBA1+ cells increase specifically within the ONL at 3, 14, and 21 DPMM, which reflects the changes seen in the Western blot (*Figure 3.2*). In contrast to the morphology seen in the inner layers of the OB where IBA1+ microglia appear with long thin processes, indicating resting microglia, the IBA1+ cells in the ONL display an amoeboid morphology consistent with an activated microglia phenotype. Additionally, the greater density of IBA1+ cells within the ONL appears to

have resolved by 40 DPMM, which is consistent with the Western blot findings, as well

as our previous study which indicated recovery of synaptic markers and olfactory

behavior by 40 DPMM (Chapter 2).

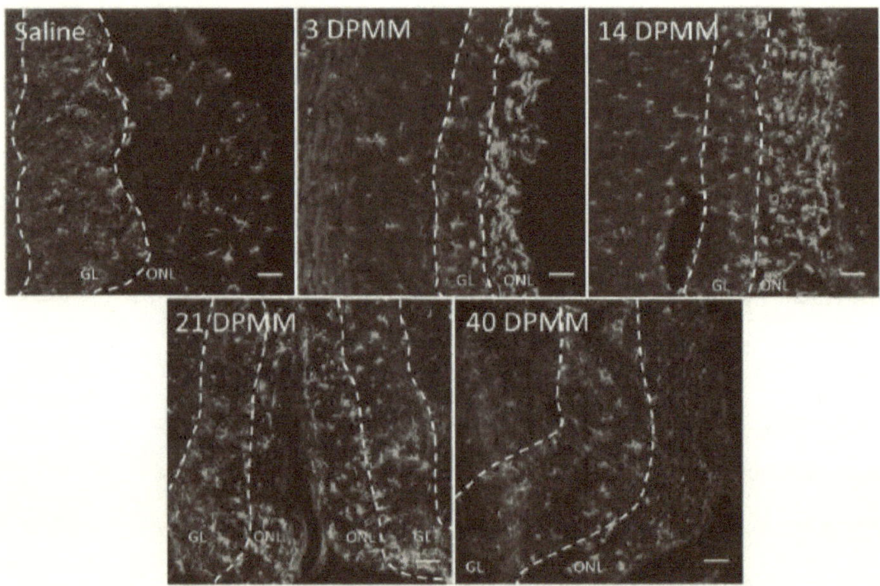

Figure 3.2. Immunofluorescent staining of IBA1+ cells (green) in the OB post-injury. GL-

glomerular layer, ONL-olfactory nerve layer. Scalebar- 50 µm.

Microglia display MGC phenotype after injury

While examining the IBA1 staining post-injury in the OB, it was noted that there

were some cells in the ONL that displayed very large cell bodies, and some had a

fainter staining pattern compared to others (*Figure 3.3A-B*). Higher magnification

imaging revealed that some of these large cells contained multiple nuclei (*Figure 3.3C-*

E). These observations are consistent with what has been previously reported as multinucleated giant cells (MGCs) (Hornik et al. 2014). MGCs are reported to be a highly phagocytic form of either microglia or macrophages (Hornik et al. 2014; Milde et al. 2015) and thus it is possible that this cell type is formed due to the massive amount of debris created within the OB after chemical ablation of the OSNs. There are conflicting reports of how MGCs are formed. It has been reported that MGCs form due to fusion of cells (Helming and Gordon 2009). However, it has also been shown that MGCs can be formed by inhibition of cytokinesis (Hornik et al. 2014). While it is difficult to conclusively demonstrate this process *in vivo*, we did find an example of an IBA1+ cell undergoing nuclear division in the GL immediately proximal to the ONL at 3 DPMM (*Figure 3.4*), suggesting that the MGCs seen in the OB are formed due to an inhibition of cytokinesis.

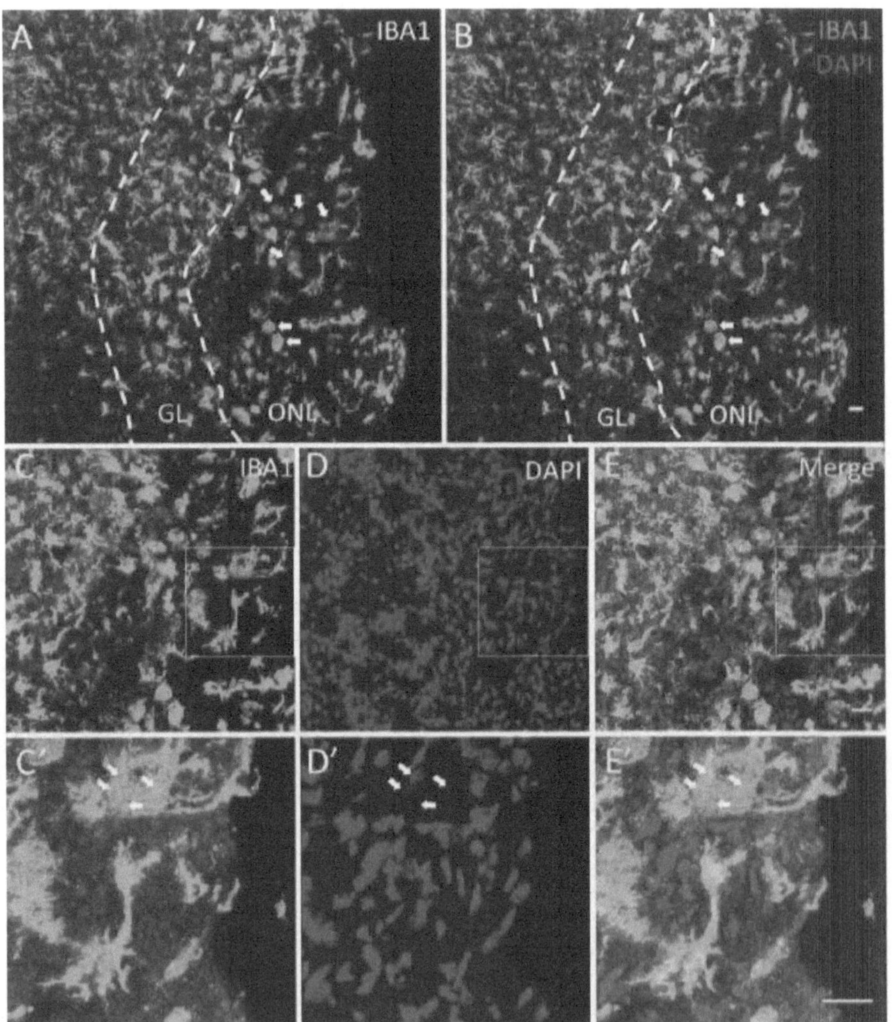

Figure 3.3. Immunofluorescent staining showing IBA+ cells in the ONL of the OB

showing multinucleated giant cell morphology at 14DPMM. A-B: Cross-section of

14DPMM OB shows large IBA1+ cells (white arrows) in the ONL. C-E: Higher mag images show an example of one of these large cells with multiple nuclei (white arrows). Scalebar-20 μm.

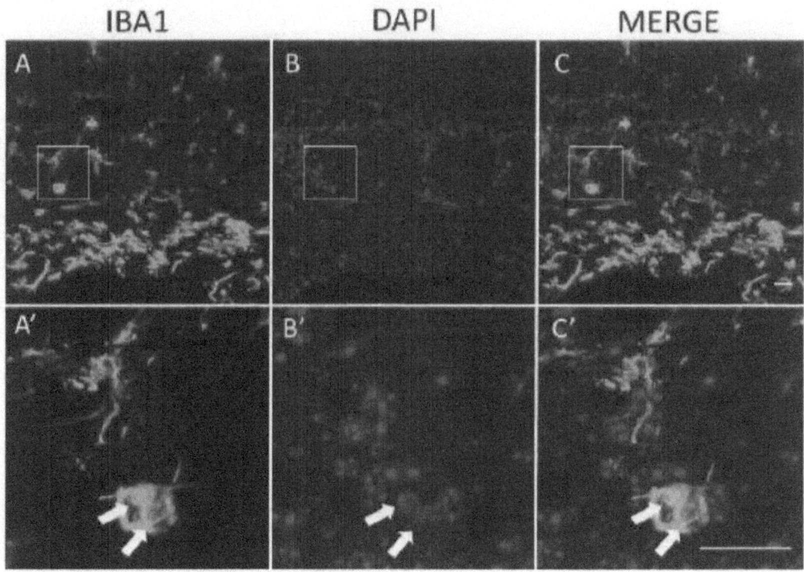

Figure 3.4. OB tissue from 3 DPMM showing an IBA1+ cell undergoing nuclear division. Scalebar- 20 μm

Phagocytosis of labeled OSN axons by microglia

In order to examine phagocytosis of axonal debris, we used a fate-tracing approach using our Cre-LoxP model used in our previous study (Chapter 2). After inducing expression of the fluorescent reporter, axons were allowed 10 days to grow to the OB and then an injection of either MM or saline was given. Tissue was collected at 3-, 7-, 14-, 21-, and 40-days post-MM/saline injection. Tissue was then stained with an

IBA1 antibody and confocal imaging was performed to determine if the IBA1+ cells in the ONL contained fluorescently labeled axonal debris. We were able to detect axonal debris within IBA1+ cells as early at 7DPMM (not shown), and also at 14DPMM (*Figure 3.5A-B*) and 21DPMM (*Figure 3.5C-D*). Our imaging did not show any debris within IBA1+ cells at 40DPMM (not shown) and fluorescently labeled axons were largely absent from the ONL. It is important to note that the fluorescently labeled axons (red) were born 24 (*Figure 3.5A-B*) and 31 (*Figure 3.5C-D*) days before the tissue was collected and have been dead for 14 (*Figure 3.5A-B*) and 21 (*Figure 3.5C-D*) days.

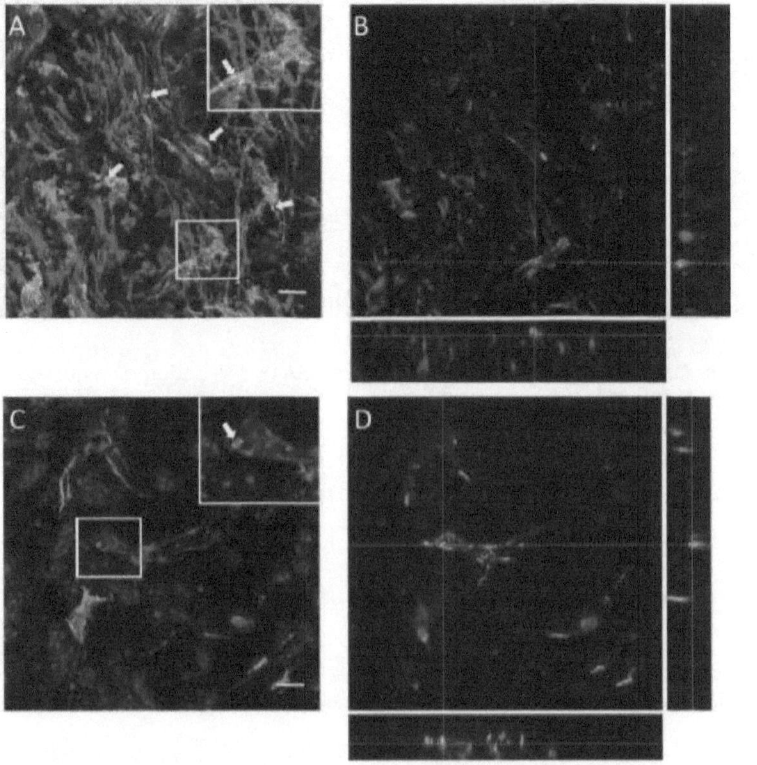

Figure 3.5. IBA1+ cells (green) phagocytose fluorescently labeled axons (red). *A:* Projection image of IBA1+ cells at 14DPMM phagocytosing labeled axons (arrows indicate yellow areas indicating colocalization). *B:* Cross-sectional analysis of the same image shows the axon indicated is contained within the IBA+ cell. *C:* Projection image of an IBA1+ cell at 21DPMM with a large piece of fluorescently labeled debris inside the cell body. *D:* Cross sectional analysis of the same image shows the piece of labeled debris is contained inside the IBA1+ cell. Scalebar- 10 μm.

Molecular mechanisms of phagocytosis

The molecular pathway underlying phagocytosis in the mammalian olfactory system after injury has not yet been defined. However, a molecular pathway has been described in ensheathing glial cells for the engulfment of axonal debris in *D. melanogaster* (Doherty et al. 2009; Doherty et al. 2014; Musashe et al. 2016). This pathway involves the receptor Draper, which has a mammalian homolog Jedi-1 that has been characterized to be involved in the engulfment process of phagocytosis, along with another receptor, Megf10, and an adaptor protein ,GULP, both *in vitro* and in glial precursor cells (Wu et al. 2009; Sullivan et al. 2014). So, we utilized qPCR to analyze expression of Jedi-1, GULP, and Megf10 in the OB after MM-induced injury (*Figure 3.6*). Our results show that expression levels of Jedi-1 and GULP after injury are not significantly altered at 3-, 14-, and 21-days post-injury. We did see a significant change in Megf10 expression at these times. However, this change was a significant decrease in Megf10 expression. At 40 DPMM, neither Jedi-1 nor Megf10 were significantly different from saline treated subjects. GULP, however, shows a significant increase at

40 DPMM, which could indicate a late stage reaction within the OB that occurs after other markers of the injury seem to have recovered.

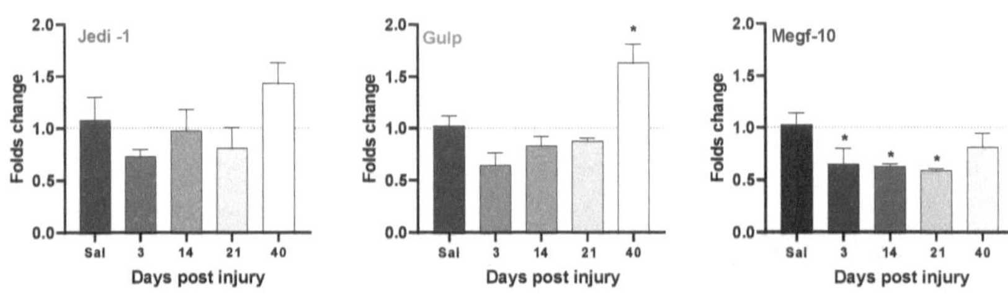

Figure 3.6. qPCR quantification of Jedi-1, GULP, and Megf10 after injury. (n=5, * p≤0.05).

Another receptor that has been shown to be play a role in phagocytosis of cellular debris is CD11b (integrin αM subunit). CD11b is also expressed on microglia, which we have shown infiltrate the ONL after a chemical ablation injury. So, we analyzed expression of CD11b after MM-induced injury by qPCR and found an increase at 3DPMM that was not statistically significant. We did, however, find a significant increase at 14- and 21DPMM. At 40DPMM, expression of CD11b had returned to control levels (*Figure 3.7*).

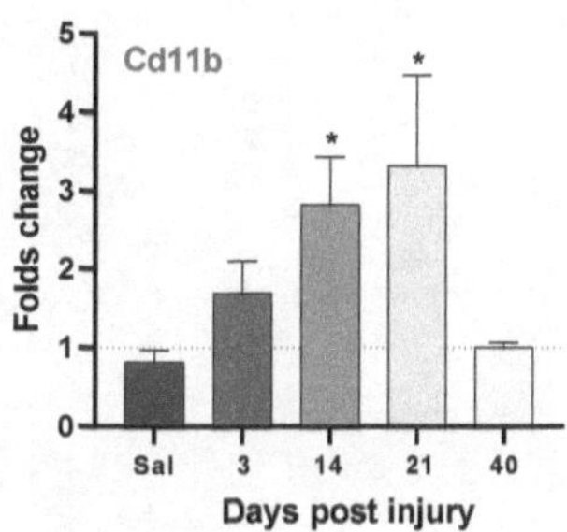

Figure 3.7. qPCR quantification of CD11b. (n=5, * p≤0.05)

To verify the expression of CD11b in microglia within the OB, we co-stained CD11b with IBA1. Immunostaining showed very little CD11b in saline treated OBs but CD11b did colocalize well with IBA1 in the ONL (*Figure 3.8A*). At 3- and 14 DPMM, however, there was a clear upregulation of CD11b in the ONL, which mimicked the upregulation of IBA1 after injury (*Figure 3.8B-C*). Higher magnification imaging at 14DPMM shows a very clear colocalization of CD11b and IBA1 in the ONL and can be seen on IBA1+ cells with multiple nuclei (*Figure 3.8D-G*).

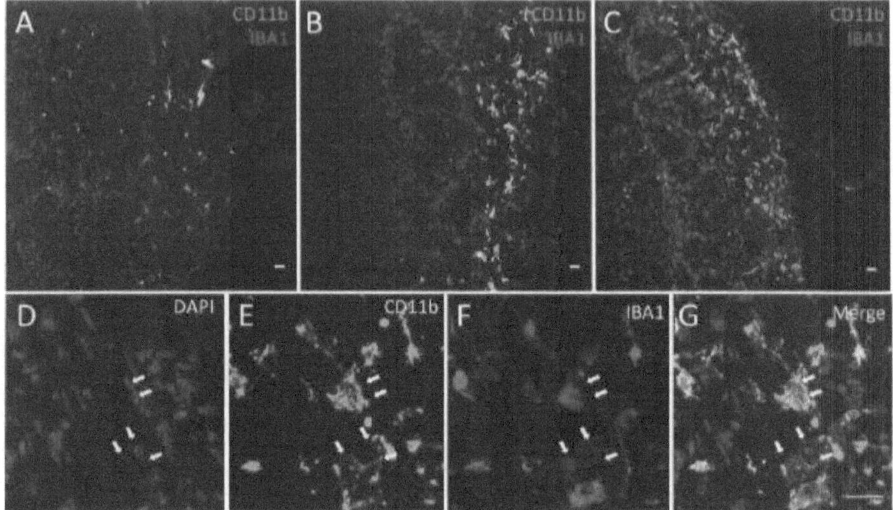

Figure 3.8. Immunofluorescent staining of IBA1 (Red) and CD11b (Green) in the OB. A: Saline. B: 3DPMM. C: 14DPMM. D-G: High magnification images show IBA1+ cells with multiple nuclei (white arrows) also expressing CD11b. Scalebar-20µm.

CD11b, also known as the alpha subunit of complement receptor 3, has been shown to play a role in complement mediated phagocytosis (Van Strijp et al. 1993). Furthermore, MGCs have been shown to be specialized in the phagocytosis of complement coated debris (Milde et al. 2015). So, with the upregulation of CD11b after injury and the appearance of an MGC morphology in the ONL, we decided to analyze components of the complement pathway to explore a possible mechanism for the removal of OSN axons after injury. We analyzed the expression of two components of complement as well as a second integrin receptor that has been shown to be involved in complement-mediated phagocytosis. Complement component C3b has been shown to bind to CD11b and is produced, along with C3a, by proteolytic cleavage of the complement 3 protein that is translated from a single mRNA for complement 3 (Comp3). Our results show an upregulation in Comp3 that mimics the trend of CD11b post injury with an increase (albeit not statistically significant) at 3DPMM and a larger, statistically significant increase at 14- and 21 DPMM (*Figure 3.9A*-note scale break in graph). Comp3 does appear to still be increased at 40 DPMM but this increase is no longer statistically significant (*Figure 3.9A*) It is possible with a larger number of samples or an analysis of protein expression would show that Comp3 is still significantly elevated at 40DPMM. An additional component of the complement pathway is C1qa, which has been shown to be produced by microglia and plays a role in synaptic pruning (Fonseca et al. 2017). Our results do seem to indicate an elevation in this mRNA, but the changes seen at 3-, 14-, and 21 DPMM are not statistically significant. The last component in our analysis of the complement mechanism was another integrin receptor, CD11c (integrin αX, alpha subunit of complement receptor 4), which has also been shown to play a role in complement mediated phagocytosis (Lukácsi et al. 2017). CD11c shows no change at 3DPMM but a significant increase at 14- and 21DPMM (*Figure 3.9C*), again showing upregulation of components of the complement pathway at these timepoints. C1qa and CD11c have not yet been analyzed at 40DPMM, but expression of both of these mRNAs at 40DPMM will be analyzed soon.

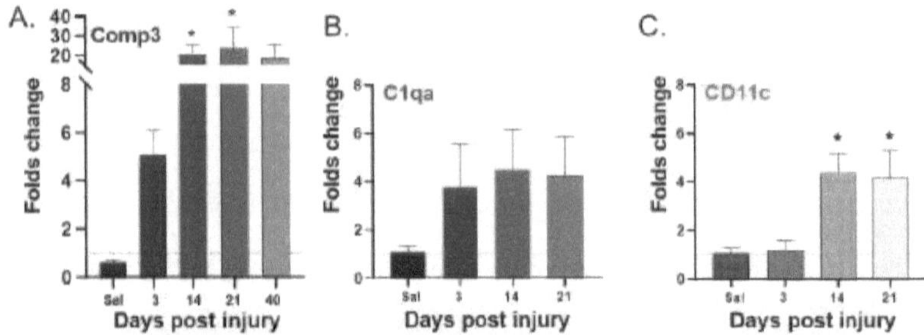

Figure 3.9. qPCR quantification of Comp 3 (*A*- Note break in scale), C1qa (*B*), and CD11c (*C*) after injury. (n=3-5, * p≤0.05)

Finally, an additional class of receptor that is classically associated with the innate immune response is the toll-like receptor (TLR). TLRs have also been shown in the central nervous system to play a role in neurodegeneration and injury response (Kielian 2006) and are associated with the activation of phagocytic mechanisms (Doyle et al. 2004). So, we analyzed expression of two TLRs, TLR2 and TLR4, in the OB post-injury via qPCR. TLR2 shows a significant increase in expression at 3DPMM but all other timepoints show no significant difference from saline treated controls (*Figure 3.10A*). TLR4, however, showed no significant change after injury when compared to controls (*Figure 3.10B*).

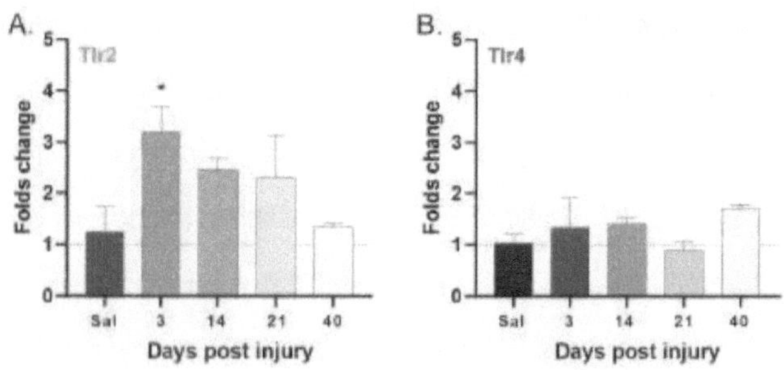

Figure 3.10. QPCR quantification of (*A*) TLR2 and (*B*) TLR4 expression post-injury. (n=5, * p≤0.05)

Discussion

The findings of this study implicate a role of microglia in removal of axonal debris after a MM-induced injury. While OECs have been suggested to be the primary phagocytes of the olfactory bulb (Suzuki et al. 1996; Su et al. 2013), our results indicate that influx of IBA1+ cells and expression of molecular markers correlated with phagocytosis suggest that microglia may also be involved after a large injury. This, however, does not necessarily mean that the OECs do not also participate in the phagocytic process after an injury but instead may suggest that the microglia may be needed in response to the large amount of neuronal debris created after a chemical ablation of the OE.

As previously mentioned, Draper has been shown in *D. melanogaster* to be involved in the phagocytic process of axonal debris (Doherty et al. 2009; Musashe et al. 2016). Our results regarding the mammalian homolog, Jedi-1, along with other associated components of this phagocytic pathway, indicate either no change or a down regulation of this pathway. The Draper pathway is associated with ensheathing glia in the *D. melanogaster* model. The lack of change or downregulation we see in our mouse model could indicate that phagocytosis by OECs does not increase due to the influx of microglia. At 40 DPMM, Jedi-1 and Megf10 are once again at control levels of expression and GULP has a significant increase. This could suggest that as the microglia influx has dissipated as shown by our IBA1 staining, that OECs once again take on primary responsibility of phagocyte in the olfactory bulb after the bulk of the debris has been removed after injury. Therefore, we hypothesize that both cell types are associated with axon debris removal, with microglia being recruited after larger insults.

The appearance of the MGC phenotype in the ONL was an unexpected finding that could have important implications in how the OB is able to remove a huge amount of debris and restore the complex synaptic network after injury. MGCs have been shown to specialize in removal of complement coated debris (Milde et al. 2015) and our results indicating upregulation of both receptors (CD11b and CD11c) and complement components (Comp3 and C1qa) could suggest a mechanism by which MGCs phagocytose axonal debris in the OB. However, the upregulation of these receptors and components do not necessarily define a full mechanism nor implicate solely the MGCs or microglia in this process. Further studies will be required to fully explore this mechanism and also determine the origin of the IBA1+ cells seen in the ONL to verify that these are microglia and not infiltrating macrophages from the periphery.

Another finding with potentially impactful implications was the early upregulation of TLR2 at 3DPMM. This was the only marker analyzed that was upregulated significantly at this timepoint. TLRs are known to activate phagocytic pathways (Doyle et al. 2004) and this could be a key receptor for activating the response to the large

amount of axonal debris created after an injury. Furthermore, TLR2 has been correlated

with the release of anti-inflammatory cytokines (Mckimmie et al. 2009) and this early

upregulation may play a role in modulating the inflammatory response in the OB that

prevents a long term inflammation and thus allows for recovery of the synaptic network.

While there are still further studies required to fully understand the role of

microglia in the response to a MM-induced injury in the OB, the findings in this study

could play a key role in our understanding of the removal of axonal debris after injury.

Furthermore, understanding the role of both the OEC and microglia in this process and

how the two interact may provide key insights into the efficiency of the OB's removal of

debris and ability to recover. The ability of the olfactory system to recover after an injury

and reestablish synaptic connections is unparalleled in other areas of the CNS and

understanding further how this is able to occur can lead to the discovery of mechanisms

that are translatable to injuries in other areas of the nervous system.

References

Blanco-Hernández E, Valle-Leija P, Zomosa-Signoret V, Drucker-Colín R, Vidaltamayo R. 2012. Odor Memory Stability after Reinnervation of the Olfactory Bulb. PLoS One. 7(10). doi:10.1371/journal.pone.0046338.

Doherty J, Logan MA, Taşdemir ÖE, Freeman MR. 2009. Ensheathing glia function as phagocytes in the adult Drosophila brain. J Neurosci. 29(15):4768–4781. doi:10.1523/JNEUROSCI.5951-08.2009.

Doherty J, Sheehan AE, Bradshaw R, Fox AN, Lu TY, Freeman MR. 2014. PI3K Signaling and Stat92E Converge to Modulate Glial Responsiveness to Axonal Injury. PLoS Biol. 12(11). doi:10.1371/journal.pbio.1001985.

Doyle SE, O'Connell RM, Miranda GA, Vaidya SA, Chow EK, Liu PT, Suzuki S, Suzuki N, Modlin RL, Yeh WC, et al. 2004. Toll-like Receptors Induce a Phagocytic

Gene Program through p38. J Exp Med. 199(1):81–90.
doi:10.1084/jem.20031237. [accessed 2020 Jul 1].
/pmc/articles/PMC1887723/?report=abstract.

Fonseca MI, Chu SH, Hernandez MX, Fang MJ, Modarresi L, Selvan P, MacGregor GR,
Tenner AJ. 2017. Cell-specific deletion of C1qa identifies microglia as the
dominant source of C1q in mouse brain. J Neuroinflammation. 14(1):48.
doi:10.1186/s12974-017-0814-9. [accessed 2020 Feb 28].
http://jneuroinflammation.biomedcentral.com/articles/10.1186/s12974-017-0814-
9.

Helming L, Gordon S. 2009. Molecular mediators of macrophage fusion. Trends Cell
Biol. 19(10):514–522. doi:10.1016/j.tcb.2009.07.005. [accessed 2020 Jun 30].
https://pubmed.ncbi.nlm.nih.gov/19733078/.

Herculano-Houzel S, Mota B, Lent R. 2006. Cellular scaling rules for rodent brains. Proc
Natl Acad Sci U S A. 103(32):12138–12143. doi:10.1073/pnas.0604911103.

Hornik TC, Neniskyte U, Brown GC. 2014. Inflammation induces multinucleation of
Microglia via PKC inhibition of cytokinesis, generating highly phagocytic
multinucleated giant cells. J Neurochem. 128(5):650–661. doi:10.1111/jnc.12477.

Kawagishi K, Ando M, Yokouchi K, Sumitomo N, Karasawa M, Fukushima N, Moriizumi
T. 2014. Stereological quantification of olfactory receptor neurons in mice.
Neuroscience. 272:29–33. doi:10.1016/j.neuroscience.2014.04.050.

Kielian T. 2006. Toll-like receptors in central nervous system glial inflammation and
homeostasis. J Neurosci Res. 83(5):711–730. doi:10.1002/jnr.20767. [accessed
2020 Jul 1]. /pmc/articles/PMC2440498/?report=abstract.

Lukácsi S, Nagy-Baló Z, Erdei A, Sándor N, Bajtay Z. 2017. The role of CR3
(CD11b/CD18) and CR4 (CD11c/CD18) in complement-mediated phagocytosis
and podosome formation by human phagocytes. Immunol Lett. 189:64–72.
doi:10.1016/j.imlet.2017.05.014.

Mckimmie CS, Moore M, Fraser AR, Jamieson T, Xu D, Burt C, Pitman NI, Nibbs RJ,
Mcinnes IB, Liew FY, et al. 2009. A TLR2 ligand suppresses inflammation by
modulation of chemokine receptors and redirection of leukocyte migration. Blood.
113(18):4224–4232. doi:10.1182/blood-2008-08-174698.The.

Milde R, Ritter J, Tennent GA, Loesch A, Martinez FO, Gordon S, Pepys MB, Verschoor
A, Helming L. 2015. Multinucleated Giant Cells Are Specialized for Complement-
Mediated Phagocytosis and Large Target Destruction. Cell Rep. 13(9):1937–
1948. doi:10.1016/j.celrep.2015.10.065. [accessed 2020 Jun 30].
/pmc/articles/PMC4675895/?report=abstract.

Musashe DT, Purice MD, Speese SD, Doherty J, Logan MA. 2016. Insulin-like Signaling
Promotes Glial Phagocytic Clearance of Degenerating Axons through Regulation

of Draper. Cell Rep. 16(7):1838–1850. doi:10.1016/j.celrep.2016.07.022. [accessed 2020 Feb 14]. http://dx.doi.org/10.1016/j.celrep.2016.07.022.

Nazareth L, Lineburg KE, Chuah MI, Tello Velasquez J, Chehrehasa F, St John JA, Ekberg JAK. 2015. Olfactory ensheathing cells are the main phagocytic cells that remove axon debris during early development of the olfactory system. J Comp Neurol. 523(3):479–494. doi:10.1002/cne.23694.

Schwob JE. 2002. Neural regeneration and the peripheral olfactory system. Anat Rec. 269(1):33–49. doi:10.1002/ar.10047. [accessed 2020 Apr 15]. http://doi.wiley.com/10.1002/ar.10047.

Schwob JE, Costanzo RM. 2010. Regeneration of the Olfactory Epithelium. Senses A Compr Ref. 4:591–612. doi:10.1016/B978-012370880-9.00115-8.

Smithson LJ, Kawaja MD. 2010. Microglial/macrophage cells in mammalian olfactory nerve fascicles. J Neurosci Res. 88(4):858–865. doi:10.1002/jnr.22254.

Van Strijp JA, Russell DG, Tuomanen E, Brown EJ, Wright SD. 1993. Ligand specificity of purified complement receptor type three (CD11b/CD18, alpha m beta 2, Mac-1). Indirect effects of an Arg-Gly-Asp (RGD) sequence. J Immunol. 151(6).

Su Z, Chen J, Qiu Y, Yuan Y, Zhu F, Zhu Y, Liu X, Pu Y, He C. 2013. Olfactory ensheathing cells: The primary innate immunocytes in the olfactory pathway to engulf apoptotic olfactory nerve debris. Glia. 61(4):490–503. doi:10.1002/glia.22450.

Sullivan CS, Scheib JL, Ma Z, Dang RP, Schafer JM, Hickman FE, Brodsky FM, Ravichandran KS, Carter BD. 2014. The adaptor protein GULP promotes Jedi-1-mediated phagocytosis through a clathrin-dependent mechanism. Mol Biol Cell. 25(12):1925–1936. doi:10.1091/mbc.E13-11-0658.

Suzuki Y, Takeda M, Farbman AI. 1996. Supporting cells as phagocytes in the olfactory epithelium after bulbectomy. J Comp Neurol. 376(4):509–517. doi:10.1002/(SICI)1096-9861(19961223)376:4<509::AID-CNE1>3.0.CO;2-5.

Wu HH, Bellmunt E, Scheib JL, Venegas V, Burkert C, Reichardt LF, Zhou Z, Farinas I, Carter BD, Farías I, et al. 2009. Glial precursors clear sensory neuron corpses during development via Jedi-1, an engulfment receptor. Nat Neurosci. 12(12):1534–1541. doi:10.1038/nn.2446.Glial.

CHAPTER 4. CONCLUSIONS

Despite a large body of work involving the study of regeneration in the olfactory system, there are still many fundamental questions that remain unanswered. The precision by which axons find synaptic targets is daunting and many molecules have been associated with this process. There is no single molecule that is responsible for synaptic targeting but instead a variety of molecules produce a combinatorial effect that results in the axonal targeting of a specific glomerulus (Mombaerts 2006). This happens not only during development, but also throughout life in terms of both normal cell turnover and in response to an injury when new OSNs are born. To understand how axons continually grow throughout the life of an organism and find synaptic targets with such precision could be key to developing more effective treatments for neurological disorders and injuries.

The data presented in this work lays groundwork for a fundamental understanding of how OSN axons regrow after an injury. Axonal fate tracing revealed that control and post-injury axons reach specific landmarks for extension from the OE to the OB at similar times. This would suggest that the molecular machinery responsible for axonal growth is intact in a post-injury system and axons are not inhibited in their growth due to debris, glial scarring, or inflammation. These latter factors have been attributed to the lack of axonal growth and regeneration in other CNS injuries (Fitch and Silver 2008). Despite axons reaching the GL around 8-10 days, our data suggest that anatomical and functional recovery in the olfactory system may take up to 40 days. It has been reported previously that axonal targeting after injury is less accurate and axons can target other glomeruli proximal to their intended glomeruli (Holbrook et al.

2014). Our data has also shown that there is an influx of microglia into the ONL after injury and activated microglia have been shown to inhibit axonal growth (Kitayama et al. 2011). So, it is possible that an injury-induced increase in microglia can play a role in the altered axonal targeting of OSNs and a delay in recovery.

While there was a significant level of microglia infiltration into the ONL after injury, it is also important to note that the levels of microglia return to near control levels by 40DPMM. This correlates with the time we observed recovery of synaptic markers in the GL and functional behavior in the hidden cookie test. It may be that the microglia are necessary to deal with the large load of axonal debris generated from the injury, but their presence is inhibitory to accurate axonal targeting. After the microglial infiltration is resolved, the axons are no longer inhibited by the presence of the microglia and axonal targeting can regain its precision. However, further studies will be needed to assess the role of microglia and axonal targeting in the olfactory bulb and if their activated state after an injury is deleterious to glomerular targeting by newly born neurons.

Another important note regarding the resolution of microglial infiltration in the olfactory bulb is that despite the presence of axonal debris from an estimated 10 million dead OSNs (Kawagishi et al. 2014), microglial activation and neuroinflammation does not appear to become chronic in the OB after injury. Microglial activation and neuroinflammation are important topics of discussion as they are often seen as precursors to neurodegenerative disease, such as Alzheimer's disease and amyotrophic lateral sclerosis (Fendrick et al. 2007; Yoshiyama et al. 2007). Furthermore, microglial activation and increase expression levels of the complement component C1q have been associated with a neurodegenerative condition involving the visual axis after a retinal injury (Silverman et al. 2016). However, despite the injury suffered by the OB, our results thus far seem to indicate a resolution of microglial activation and seems to suggest that the OB does not enter into a state of chronic inflammation after injury. The OB seems to have mechanisms in place that aid in dealing with inflammation that have not yet been defined. In a study designed to mimic chronic nasal inflammation by repeated intranasal injection of lipopolysaccharide (LPS), the OB was shown to be atrophic during the chronic inflammation but was able to return to levels consistent with the OB prior to LPS treatment upon cessation of the injections (Hasegawa-Ishii et al. 2019). Receptors involved in mediation of inflammatory responses would be of great interest in defining how the OB is able to endure a large-scale injury without entering into a state of chronic neuroinflammation. While our results do seem to reflect that the OB is able to undergo an injury without displaying signs of neuroinflammation in the time course of our study, it would be useful to assess the long-term condition of the olfactory bulb post-injury as compared to age-matched controls to determine if there are long-term detriments caused from a chemical ablation injury. Furthermore, it would also be valuable to explore susceptibility and response to further injury.

While our results regarding phagocytosis focused on the role of microglia, the role of the OEC in this process should not be overlooked. The OEC has previously been shown to be involved in phagocytosis in the OB (Su et al. 2013) and the influx of microglia does not necessarily mean that the OECs don't contribute to the removal of axonal debris after injury. Our results show that Jedi-1, GULP, and Megf10 are either unchanged or downregulated at the times when we see infiltration of microglia in the

ONL. Based on the Jedi-1 homology with Draper and its association with ensheathing glia in *D. melanogaster* (Doherty et al. 2009; Doherty et al. 2014; Musashe et al. 2016), we hypothesis that these three markers are expressed on the mammalian OEC and may underly phagocytosis during normal regeneration. However, their expression on OECs would need to be confirmed with immunohistochemical staining and their role in phagocytosis during normal regeneration would require further study.

While it is plausible that Jedi-1, GULP, and Megf10 are the molecular pathway underlying OEC phagocytosis and microglia phagocytose axonal debris via complement, there are multiple other pathways that have been described to be involved in phagocytosis (Sierra et al. 2013). Of note are a group of receptor tyrosine kinases in the TAM (Tyro3, Axl, and Mertk) family that have been associated with microglial phagocytosis and the deletion of these receptors leads to increased inflammation (Grommes et al. 2008; Weinger et al. 2011). It would be valuable analyze other receptors involved in phagocytosis, such as the TAM family, to see if other receptors may be involved in the phagocytic process. CD11b has been described in functions other than phagocytosis, such as adhesion and migration (Zen et al. 2011; Hyun et al. 2019). Therefore, the upregulation of CD11b seen in our experiments does not necessarily conclude that CD11b is the receptor underlying phagocytosis. The increase in expression and colocalization with microglia does, however, warrant further investigation into the role of CD11b in microglia response to injury in the OB.

While the OEC's role in post-injury debris removal is undetermined at this point, OECs do have other roles that make them of further interest in regeneration. There is great interest in cellular transplantation using the OEC in injuries such as spinal cord injuries (Mayeur et al. 2013; Gómez et al. 2018). The OEC is believed to allow axonal growth in the presence of a glial scar, which again is often attributed to the inhibition of axonal growth in the CNS (Moreno-Flores et al. 2002; Fitch and Silver 2008). As there is a large increase in microglia within the ONL after injury and it has been established that microglia can inhibit axonal growth (Kitayama et al. 2011), it would be valuable to determine if OECs are critical for the growth of new axons among the influx of microglia in the OB and what mechanism OECs employ to create an environment that is conducive to axon growth. A greater understanding of the molecular mechanisms that OECs employ to allow for axon growth would have great value in translation to treatments of other neural injuries.

Finally, the results of this study and others using this fate tracing method for mapping regrowth of axons (Rodriguez-Gil et al. 2015; Liberia et al. 2019), have established distinct timelines for axonal growth. The spatiotemporal resolution of this technique can be utilized in conjunction with more mechanistic approaches, such as using in conjunction with RNA sequencing, to uncover mechanisms responsible for certain aspects of axon growth and axon guidance with a precision that has not been produced to this point. Correlating axon growth landmarks with the onset of expression of certain genes known to be involved in axon guidance could begin to unravel the large puzzle that is axon guidance in the olfactory system.

REFERENCES

Albeanu DF, Provost AC, Agarwal P, Soucy ER, Zak JD, Murthy VN. 2018. Olfactory marker protein (OMP) regulates formation and refinement of the olfactory glomerular map. Nat Commun. 9(1):1–12. doi:10.1038/s41467-018-07544-9.

Alenius M, Bohm S. 1997. Identification of a novel neural cell adhesion molecule-related gene with a potential role in selective axonal projection. J Biol Chem. 272(42):26083–26086. doi:10.1074/jbc.272.42.26083.

Arandjelovic S, Ravichandran KS. 2015. Phagocytosis of apoptotic cells in homeostasis. Nat Immunol. 16(9):907–917. doi:10.1038/ni.3253.

Ashwell K. 2012. The Olfactory System. In: The Mouse Nervous System. Elsevier Inc. p. 653–660.

Astic L, Pellier-Monnin V, Saucier D, Charrier C, Mehlen P. 2002. Expression of netrin-1 and netrin-1 receptor, DCC, in the rat olfactory nerve pathway during development and axonal regeneration. Neuroscience. 109(4):643–656. doi:10.1016/S0306-4522(01)00535-8. [accessed 2019 Sep 7]. https://www.sciencedirect.com/science/article/abs/pii/S0306452201005358?via% 3Dihub.

Astic L, Saucier D. 1986. Anatomical mapping of the neuroepithelial projection to the olfactory bulb in the rat. Brain Res Bull. 16(4):445–454. doi:10.1016/0361-9230(86)90172-3.

Au E, Roskams AJ. 2003. Olfactory ensheathing cells of the lamina propria in vivo and in vitro. Glia. 41(3):224–236. doi:10.1002/glia.10160.

Au WW, Treloar HB, Greer CA. 2002. Sublaminar organization of the mouse olfactory bulb nerve layer. J Comp Neurol. 446(1):68–80. doi:10.1002/cne.10182. [accessed 2020 Feb 15]. http://doi.wiley.com/10.1002/cne.10182.

Baker H, Kawano T, Margolis FL, Joh TH. 1983. TRANSNEURONAL REGULATION OF TYROSINE HYDROXYLASE EXPRESSION IN OLFACTORY BULB OF MOUSE AND RAT1. J Neurosci. 3(1):69–78.

Barber PC, Dahl D. 1987. Glial fibrillary acidic protein (GFAP)-like immunoreactivity in normal and transected rat olfactory nerve. Exp Brain Res. 65(3):681–685. doi:10.1007/BF00235993.

Barber PC, Lindsay RM. 1982. Schwann cells of the olfactory nerves contain glial fibrillary acidic protein and resemble astrocytes. Neuroscience. 7(12):3077–3090. doi:10.1016/0306-4522(82)90231-7.

Barraud P, St John JA, Stolt CC, Wegner M, Baker CVH. 2013. Olfactory ensheathing glia are required for embryonic olfactory axon targeting and the migration of gonadotropin-releasing hormone neurons. Biol Open. 2(7):750–759. doi:10.1242/bio.20135249.

Bhattacharyya A, Oppenheim RW, Prevette D, Moore BW, Brackenbury R, Ratner N. 1992. S100 is present in developing chicken neurons and schwann cell and promotes motor neuron survival in vivo. J Neurobiol. 23(4):451–466. doi:10.1002/neu.480230410.

Blanco-Hernández E, Valle-Leija P, Zomosa-Signoret V, Drucker-Colín R, Vidaltamayo R. 2012. Odor Memory Stability after Reinnervation of the Olfactory Bulb. PLoS One. 7(10). doi:10.1371/journal.pone.0046338.

Bonfanti L. 2006. PSA-NCAM in mammalian structural plasticity and neurogenesis. Prog Neurobiol. 80(3):129–164. doi:10.1016/j.pneurobio.2006.08.003.

Booker-Dwyer T, Hirsh S, Zhao H. 2008. A unique cell population in the mouse olfactory bulb displays nuclear β-catenin signaling during development and olfactory sensory neuron regeneration. Dev Neurobiol. 68(7):859–869. doi:10.1002/dneu.20606. [accessed 2019 Sep 7]. http://doi.wiley.com/10.1002/dneu.20606.

Brandt I, Brittebo EB, Feil VJ, Bakke JE. 1990. Irreversible binding and toxicity of the herbicide dichlobenil (2,6-dichlorobenzonitrile) in the olfactory mucosa of mice. Toxicol Appl Pharmacol. 103(3):491–501. doi:10.1016/0041-008X(90)90322-L.

Brann JH, Firestein SJ. 2014. A lifetime of neurogenesis in the olfactory system. Front Neurosci. 8:182. doi:10.3389/fnins.2014.00182. [accessed 2019 Sep 16]. http://journal.frontiersin.org/article/10.3389/fnins.2014.00182/abstract.

Brown GC, Neher JJ. 2014. Microglial phagocytosis of live neurons. Nat Rev Neurosci. 15(4):209–216. doi:10.1038/nrn3710.

Bulfone A, Wang F, Hevner R, Anderson S, Cutforth T, Chen S, Meneses J, Pedersen R, Axel R, Rubenstein JLR. 1998. An olfactory sensory map develops in the absence of normal projection neurons or GABAergic interneurons. Neuron. 21(6):1273–1282. doi:10.1016/S0896-6273(00)80647-9.

Burd GD. 1993. Morphological study of the effects of intranasal zinc sulfate irrigation on the mouse olfactory epithelium and olfactory bulb. Microsc Res Tech. 24(3):195–213. doi:10.1002/jemt.1070240302.

Burek MJ, Oppenheim RW. 1996. Programmed cell death in the developing nervous system. In: Brain Pathology. Vol. 6. Blackwell Publishing Ltd. p. 427–446.

Caggiano M, Kauer JS, Hunter DD. 1994. Globose Basal Cells Are Neuronal Progenitors in the Olfactory Epithelium: A Lineage Analysis Using a Replication-Incompetent Retrovirus. Neuron. 13(2):339–352. doi:10.1016/0896-6273(94)90351-4.

Cai YJ, Wang F, Chen ZX, Li L, Fan H, Wu ZB, Ge JF, Hu W, Wang QN, Zhu DF. 2018. Hashimoto's thyroiditis induces neuroinflammation and emotional alterations in euthyroid mice. J Neuroinflammation. 15(299). doi:10.1186/s12974-018-1341-z.

Carr VM, Walters E, Margolis FL, Farbman AI. 1998. An enhanced olfactory marker protein immunoreactivity in individual olfactory receptor neurons following olfactory bulbectomy may be related to increased neurogenesis. J Neurobiol. 34(4):377–90. [accessed 2019 Sep 16]. http://www.ncbi.nlm.nih.gov/pubmed/9514526.

Cau E, Casarosa S, Guillemot F. 2002. Mash1 and Ngn1 control distinct steps of determination and differentiation in the olfactory sensory neuron lineage. Development. 129(8):1871–1880.

Chen X, Fang H, Schwob JE. 2004. Multipotency of Purified, Transplanted Globose Basal Cells in Olfactory Epithelium. J Comp Neurol. 469(4):457–474. doi:10.1002/cne.11031.

Chesler AT, Zou D-J, Le Pichon CE, Peterlin ZA, Matthews GA, Pei X, Miller MC, Firestein S. 2007. A G protein/cAMP signal cascade is required for axonal convergence into olfactory glomeruli. Proc Natl Acad Sci. 104(3):1039–1044. doi:10.1073/pnas.0609215104.

Chess A, Simon I, Cedar H, Axel R. 1994. Allelic inactivation regulates olfactory receptor gene expression. Cell. 78(5):823–34. doi:10.1016/s0092-8674(94)90562-2. [accessed 2019 Sep 7]. http://www.ncbi.nlm.nih.gov/pubmed/8087849.

Cho JH, Kam JWK, Cloutier JF. 2012. Slits and Robo-2 regulate the coalescence of subsets of olfactory sensory neuron axons within the ventral region of the olfactory bulb. Dev Biol. 371(2):269–279. doi:10.1016/j.ydbio.2012.08.028.

Cho JH, Lépine M, Andrews W, Parnavelas J, Cloutier JF. 2007. Requirement for slit-1 and robo-2 in zonal segregation of olfactory sensory neuron axons in the main olfactory bulb. J Neurosci. 27(34):9094–9104. doi:10.1523/JNEUROSCI.2217-07.2007.

Coleman JH, Lin B, Louie JD, Peterson J, Lane RP, Schwob JE. 2019. Spatial Determination of Neuronal Diversification in the Olfactory Epithelium. J Neurosci. 39(5):814–832. doi:10.1523/JNEUROSCI.3594-17.2018. [accessed 2020 Feb 1]. https://doi.org/10.1523/JNEUROSCI.3594-17.2018.

Costanzo RM. 1985. Neural regeneration and functional reconnection following olfactory nerve transection in hamster. Brain Res. 361(1–2):258–266. doi:10.1016/0006-8993(85)91297-1.

Costanzo RM, Graziadei PPC. 1983. A quantitative analysis of changes in the olfactory epithelium following bulbectomy in hamster. J Comp Neurol. 215(4):370–381. doi:10.1002/cne.902150403.

Crandall JE, Dibble C, Butler D, Pays L, Ahmad N, Kostek C, Püschel AW, Schwarting GA. 2000. Patterning of olfactory sensory connections is mediated by extracellular matrix proteins in the nerve layer of the olfactory bulb. J Neurobiol. 45(4):195–206. [accessed 2019 Sep 9]. http://www.ncbi.nlm.nih.gov/pubmed/11077424.

Cremer H, Lange R, Christoph A, Plomann M, Vopper G, Roes J, Brown R, Baldwin S, Kraemer P, Scheff S, et al. 1994. Inactivation of the N-CAM gene in mice results in size reduction of the olfactory bulb and deficits in spatial learning. Nature. 367(6462):455–459. doi:10.1038/367455a0. [accessed 2020 Jul 15]. https://www.nature.com/articles/367455a0.

Davalos D, Grutzendler J, Yang G, Kim J V., Zuo Y, Jung S, Littman DR, Dustin ML, Gan WB. 2005. ATP mediates rapid microglial response to local brain injury in vivo. Nat Neurosci. 8(6):752–758. doi:10.1038/nn1472.

Deckner ML, Risling M, Frisén J. 1997. Apoptotic death of olfactory sensory neurons in the adult rat. Exp Neurol. 143(1):132–140. doi:10.1006/exnr.1996.6352.

DeMaria S, Ngai J. 2010. The cell biology of smell. J Cell Biol. 191(3):443–452. doi:10.1083/jcb.201008163.

Desjardins M, Huber LA, Parton RG, Griffiths G. 1994. Biogenesis of phagolysosomes proceeds through a sequential series of interactions with the endocytic apparatus. J Cell Biol. 124(5):677–688. doi:10.1083/jcb.124.5.677.

Doetsch F, Caille I, Lim DA, Garcı JM, Alvarez-buylla A. 1999. Subventricular Zone Astrocytes Are Neural Stem Cells in the Adult Mammalian Brain. Cell. 97:703–716.

Doherty J, Logan MA, Taşdemir ÖE, Freeman MR. 2009. Ensheathing glia function as phagocytes in the adult Drosophila brain. J Neurosci. 29(15):4768–4781. doi:10.1523/JNEUROSCI.5951-08.2009.

Doherty J, Sheehan AE, Bradshaw R, Fox AN, Lu TY, Freeman MR. 2014. PI3K Signaling and Stat92E Converge to Modulate Glial Responsiveness to Axonal Injury. PLoS Biol. 12(11). doi:10.1371/journal.pbio.1001985.

Doucette R. 1996. Immunohistochemical localization of laminin, fibronectin and collagen type IV in the nerve fiber layer of the olfactory bulb. Int J Dev Neurosci. 14(7–8):945–959. doi:10.1016/S0736-5748(96)00042-1. [accessed 2020 Feb 11]. http://doi.wiley.com/10.1016/S0736-5748%2896%2900042-1.

Doyle SE, O'Connell RM, Miranda GA, Vaidya SA, Chow EK, Liu PT, Suzuki S, Suzuki N, Modlin RL, Yeh WC, et al. 2004. Toll-like Receptors Induce a Phagocytic Gene Program through p38. J Exp Med. 199(1):81–90. doi:10.1084/jem.20031237. [accessed 2020 Jul 1]. /pmc/articles/PMC1887723/?report=abstract.

Van Eldik LJ, Christie-Pope B, Bolin LM, Shooter EM, Whetsell WO. 1991. Neurotrophic activity of S-100β in cultures of dorsal root ganglia from embryonic chick and fetal rat. Brain Res. 542(2):280–285. doi:10.1016/0006-8993(91)91579-P.

Fadok VA, Bratton DL, Rose DM, Pearson A, Ezekewitz RAB, Henson PM. 2000. A receptor for phosphatidylserine-specific clearance of apoptotic cells. Nature. 405(6782):85–90. doi:10.1038/35011084.

Fadok VA, Voelker DR, Campbell PA, Cohen JJ, Bratton DL, Henson PM. 1992. Exposure of phosphatidylserine on the surface of apoptotic lymphocytes triggers specific recognition and removal by macrophages. J Immunol. 148(7):2207–16. [accessed 2020 Apr 28]. http://www.ncbi.nlm.nih.gov/pubmed/1545126.

Farbman AI. 1990. Olfactory neurogenesis: genetic or environmental controls? Trends Neurosci. 13(9):362–365. doi:10.1016/0166-2236(90)90017-5. [accessed 2019

Sep 16].
https://www.sciencedirect.com/science/article/abs/pii/0166223690900175?via%3
Dihub.

Farbman AI. 1992. Cell Biology of Olfaction. Cambridge University Press.

Feinstein P, Bozza T, Rodriguez I, Vassalli A, Mombaerts P. 2004. Axon guidance of
mouse olfactory sensory neurons by odorant receptors and the β2 adrenergic
receptor. Cell. 117(6):833–846. doi:10.1016/j.cell.2004.05.013. [accessed 2019
Sep 7]. http://www.ncbi.nlm.nih.gov/pubmed/15186782.

Feinstein P, Mombaerts P. 2004. A contextual model for axonal sorting into glomeruli in
the mouse olfactory system. Cell. 117(6):817–31. doi:10.1016/j.cell.2004.05.011.
[accessed 2019 Sep 7]. http://www.ncbi.nlm.nih.gov/pubmed/15186781.

Fendrick SE, Xue QS, Streit WJ. 2007. Formation of multinucleated giant cells and
microglial degeneration in rats expressing a mutant Cu/Zn superoxide dismutase
gene. J Neuroinflammation. 4(1):9. doi:10.1186/1742-2094-4-9. [accessed 2020
Feb 28]. http://jneuroinflammation.biomedcentral.com/articles/10.1186/1742-
2094-4-9.

Fitch MT, Silver J. 2008. CNS injury, glial scars, and inflammation: Inhibitory
extracellular matrices and regeneration failure. Exp Neurol. 209(2):294–301.
doi:10.1016/j.expneurol.2007.05.014. [accessed 2020 Jul 1].
https://pubmed.ncbi.nlm.nih.gov/17617407/.

Fonseca MI, Chu SH, Hernandez MX, Fang MJ, Modarresi L, Selvan P, MacGregor GR,
Tenner AJ. 2017. Cell-specific deletion of C1qa identifies microglia as the
dominant source of C1q in mouse brain. J Neuroinflammation. 14(1):48.
doi:10.1186/s12974-017-0814-9. [accessed 2020 Feb 28].
http://jneuroinflammation.biomedcentral.com/articles/10.1186/s12974-017-0814-
9.

Fourgeaud L, Traves PG, Tufail Y, Leal-Bailey H, Lew ED, Burrola PG, Callaway P,
Zagorska A, Rothlin C V., Nimmerjahn A, et al. 2016. TAM receptors regulate
multiple features of microglial physiology. Nature. 532(7598):240–244.
doi:10.1038/nature17630.

Franceschini IA, Barnett SC. 1996. Low-affinity NGF-receptor and E-N-CAM expression
define two types of olfactory nerve ensheathing cells that share a common
lineage. Dev Biol. doi:10.1006/dbio.1996.0027.

Fricker M, Neher JJ, Zhao JW, Théry C, Tolkovsky AM, Brown GC. 2012. MFG-E8
mediates primary phagocytosis of viable neurons during neuroinflammation. J
Neurosci. 32(8):2657–2666. doi:10.1523/JNEUROSCI.4837-11.2012.

Fuller AD, Van Eldik LJ. 2008. MFG-E8 regulates microglial phagocytosis of apoptotic
neurons. J Neuroimmune Pharmacol. 3(4):246–256. doi:10.1007/s11481-008-
9118-2.

Fung KM, Peringa J, Venkatachalam S, Lee VMY, Trojanowski JQ. 1997. Coordinate reduction in cell proliferation and cell death in mouse olfactory epithelium from birth to maturity. Brain Res. 761(2):347–351. doi:10.1016/S0006-8993(97)00467-8.

Garin J, Diez R, Kieffer S, Dermine JF, Duclos S, Gagnon E, Sadoul R, Rondeau C, Desjardins M. 2001. The phagosome proteome: Insight into phagosome functions. J Cell Biol. 152(1):165–180. doi:10.1083/jcb.152.1.165.

Gehler S, Gallo G, Veien E, Letourneau PC. 2004. p75 Neurotrophin Receptor Signaling Regulates Growth Cone Filopodial Dynamics through Modulating RhoA Activity. J Neurosci. 24(18):4363–4372. doi:10.1523/JNEUROSCI.0404-04.2004.

Genter MB, Deamer NJ, Blake BL, Wesley DS, Levi PE. 1995. Olfactory Toxicity of Methimazole: Dose-Response and Structure-Activity Studies and Characterization of Flavin-Containing Monooxygenase Activity in the Long-Evans Rat Olfactory Mucosa. Toxicol Pathol. 23(4):477–486. doi:10.1177/019262339502300404. [accessed 2019 Sep 16]. http://journals.sagepub.com/doi/10.1177/019262339502300404.

Gogos JA, Osborne J, Nemes A, Mendelsohn M, Axel R. 2000. Genetic ablation and restoration of the olfactory topographic map. Cell. 103(4):609–20. doi:10.1016/s0092-8674(00)00164-1. [accessed 2019 Sep 7]. http://www.ncbi.nlm.nih.gov/pubmed/11106731.

Goldstein BJ, Schwob JE. 1996. Analysis of the globose basal cell compartment in rat olfactory epithelium using GBC-1, a new monoclonal antibody against globose basal cells. J Neurosci. 16(12):4005–4016. doi:10.1523/jneurosci.16-12-04005.1996.

Gómez RM, Sánchez MY, Portela-Lomba M, Ghotme K, Barreto GE, Sierra J, Moreno-Flores MT. 2018. Cell therapy for spinal cord injury with olfactory ensheathing glia cells (OECs). Glia. 66(7):1267–1301. doi:10.1002/glia.23282.

Graziadei PPC, Levine RR, Monti Graziadei GA. 1978. Regeneration of olfactory axons and synapse formation in the forebrain after bulbectomy in neonatal mice. Proc Natl Acad Sci U S A. 75(10):5230–5234. doi:10.1073/pnas.75.10.5230.

Graziadei PPC, Levine RR, Monti Graziadei GA. 1979. Plasticity of connections of the olfactory sensory neuron: Regeneration into the forebrain following bulbectomy in the neonatal mouse. Neuroscience. 4(6):713–727. doi:10.1016/0306-4522(79)90002-2.

Graziadei PPC, Monti Graziadei GA. 1980. Neurogenesis and neuron regeneration in the olfactory system of mammals. III. Deafferentation and reinnervation of the olfactory bulb following section of thefila olfactoria in rat. J Neurocytol. 9(2):145–162. doi:10.1007/BF01205155. [accessed 2019 Sep 16]. http://link.springer.com/10.1007/BF01205155.

Greer CA, Whitman MC, Rela L, Imamura F, Gil DR, Haven N. 2008. 4 . 36 Architecture of the Olfactory Bulb. Architecture.

Grommes C, Lee CYD, Wilkinson BL, Jiang Q, Koenigsknecht-Talboo JL, Varnum B, Landreth GE. 2008. Regulation of microglial phagocytosis and inflammatory gene expression by Gas6 acting on the Axl/Mer family of tyrosine kinases. J NeuroImmune Pharmacol. 3(2):130–140. doi:10.1007/s11481-007-9090-2. [accessed 2020 May 11]. /pmc/articles/PMC3653274/?report=abstract.

Gu J, Zhang QY, Genter MB, Lipinskas TW, Negishi M, Nebert DW, Ding X. 1998. Purification and characterization of heterologously expressed mouse CYP2A5 and CYP2G1: role in metabolic activation of acetaminophen and 2,6-dichlorobenzonitrile in mouse olfactory mucosal microsomes. J Pharmacol Exp Ther. 285(3):1287–95.

Hardy D, Saghatelyan A. 2017. Different forms of structural plasticity in the adult olfactory bulb. Neurogenesis. 4(1):e1301850. doi:10.1080/23262133.2017.1301850. [accessed 2019 Aug 25]. http://www.ncbi.nlm.nih.gov/pubmed/28596977.

Hasegawa-Ishii S, Shimada A, Imamura F. 2019. Neuroplastic changes in the olfactory bulb associated with nasal inflammation in mice. J Allergy Clin Immunol. 143(3):978-989.e3. doi:10.1016/j.jaci.2018.09.028. [accessed 2020 Jul 1]. https://doi.org/10.1016/j.jaci.2018.09.028.

Helming L, Gordon S. 2009. Molecular mediators of macrophage fusion. Trends Cell Biol. 19(10):514–522. doi:10.1016/j.tcb.2009.07.005. [accessed 2020 Jun 30]. https://pubmed.ncbi.nlm.nih.gov/19733078/.

Henion TR, Raitcheva D, Grosholz R, Biellmann F, Skarnes WC, Hennet T, Schwarting GA. 2005. Beta1,3-N-acetylglucosaminyltransferase 1 glycosylation is required for axon pathfinding by olfactory sensory neurons. J Neurosci. 25(8):1894–903. doi:10.1523/JNEUROSCI.4654-04.2005. [accessed 2019 Sep 9]. http://www.ncbi.nlm.nih.gov/pubmed/15728829.

Herbert RP, Harris J, Chong KP, Chapman J, West AK, Chuah MI. 2012. Cytokines and olfactory bulb microglia in response to bacterial challenge in the compromised primary olfactory pathway. J Neuroinflammation. 9(109). doi:10.1186/1742-2094-9-109.

Herculano-Houzel S, Mota B, Lent R. 2006. Cellular scaling rules for rodent brains. Proc Natl Acad Sci U S A. 103(32):12138–12143. doi:10.1073/pnas.0604911103.

Hökfelt T, Stanic D, Sanford SD, Gatlin JC, Nilsson I, Paratcha G, Ledda F, Fetissov S, Lindfors C, Herzog H, et al. 2008. NPY and its involvement in axon guidance, neurogenesis, and feeding. Nutrition. 24(9):860–868. doi:10.1016/j.nut.2008.06.010.

Holbrook EH, Iwema CL, Peluso CE, Schwob JE. 2014. The regeneration of P2 olfactory sensory neurons is selectively impaired following methyl bromide lesion. Chem Senses. 39(7):601–616. doi:10.1093/chemse/bju033.

Holbrook EH, Leopold DA, Schwob JE. 2005. Abnormalities of axon growth in human olfactory mucosa. Laryngoscope. 115(12):2144–2154.

doi:10.1097/01.MLG.0000181493.83661.CE.

Holcomb JD, Mumm JS, Calof AL. 1995. Apoptosis in the Neuronal Lineage of the
Mouse Olfactory Epithelium: Regulationin Vivoandin Vitro. Dev Biol. 172(1):307–
323. doi:10.1006/dbio.1995.0025.

Honda S, Sasaki Y, Ohsawa K, Imai Y, Nakamura Y, Inoue K, Kohsaka S. 2001.
Extracellular ATP or ADP induce chemotaxis of cultured microglia through Gi/o-
coupled P2Y receptors. J Neurosci. 21(6):1975–1982. doi:10.1523/jneurosci.21-
06-01975.2001.

Hornik TC, Neniskyte U, Brown GC. 2014. Inflammation induces multinucleation of
Microglia via PKC inhibition of cytokinesis, generating highly phagocytic
multinucleated giant cells. J Neurochem. 128(5):650–661. doi:10.1111/jnc.12477.

Huard JMT, Youngentob SL, Goldstein BJ, Luskin MB, Schwob JE. 1998. Adult
olfactory epithelium contains multipotent progenitors that give rise to neurons and
non-neural cells. J Comp Neurol. 400(4):469–486. doi:10.1002/(SICI)1096-
9861(19981102)400:4<469::AID-CNE3>3.0.CO;2-8.

Hurtt ME, Thomas DA, Working PK, Monticello TM, Morgan KT. 1988. Degeneration
and regeneration of the olfactory epithelium following inhalation exposure to
methyl bromide: Pathology, cell kinetics, and olfactory function. Toxicol Appl
Pharmacol. 94(2):311–328. doi:10.1016/0041-008X(88)90273-6. [accessed 2019
Sep 16].
https://www.sciencedirect.com/science/article/pii/0041008X88902736?via%3Dihu
b.

Huynh MLN, Fadok VA, Henson PM. 2002. Phosphatidylserine-dependent ingestion of
apoptotic cells promotes TGF-β1 secretion and the resolution of inflammation. J
Clin Invest. 109(1):41–50. doi:10.1172/JCI0211638.

Hyun YM, Choe YH, Park SA, Kim M. 2019. LFA-1 (CD11a/CD18) and Mac-1
(CD11b/CD18) distinctly regulate neutrophil extravasation through hotspots I and
II. Exp Mol Med. 51(4):1–13. doi:10.1038/s12276-019-0227-1. [accessed 2020
Jul 1]. https://doi.org/10.1038/s12276-019-0227-1.

Imai T, Suzuki M, Sakano H. 2006. Odorant receptor-derived cAMP signals direct
axonal targeting. Science (80-). 314(5799):657–661.
doi:10.1126/science.1131794.

Iwema CL, Fang H, Kurtz DB, Youngentob SL, Schwob JE. 2004. Odorant Receptor
Expression Patterns Are Restored in Lesion-Recovered Rat Olfactory Epithelium.
J Neurosci. 24(2):356–369. doi:10.1523/JNEUROSCI.1219-03.2004.

Jang W, Chen X, Flis D, Harris M, Schwob JE. 2014. Label-Retaining, Quiescent
Globose Basal Cells Are Found in the Olfactory Epithelium. J Comp Neurol.
522(4):731–749. doi:10.1002/cne.23470.

Jang W, Youngentob SL, Schwob JE. 2003. Globose basal cells are required for
reconstitution of olfactory epithelium after methyl bromide lesion. J Comp Neurol.

460(1):123–140. doi:10.1002/cne.10642.

St. John JA, Pasquale EB, Key B. 2002. EphA receptors and ephrin-A ligands exhibit highly regulated spatial and temporal expression patterns in the developing olfactory system. Dev Brain Res. 138(1):1–14. doi:10.1016/S0165-3806(02)00454-6. [accessed 2019 Sep 7]. https://www.sciencedirect.com/science/article/abs/pii/S0165380602004546?via%3Dihub.

Kafitz KW, Greer CA. 1997. Role of laminin in axonal extension from olfactory receptor cells. J Neurobiol. 32(3):298–310. doi:10.1002/(SICI)1097-4695(199703)32:3<298::AID-NEU4>3.0.CO;2-2. [accessed 2020 Feb 11]. http://doi.wiley.com/10.1002/%28SICI%291097-4695%28199703%2932%3A3%3C298%3A%3AAID-NEU4%3E3.0.CO%3B2-2.

Kafitz KW, Greer CA. 1998. Differential expression of extracellular matrix and cell adhesion molecules in the olfactory nerve and glomerular layers of adult rats. J Neurobiol. 34(3):271–82. [accessed 2019 Sep 7]. http://www.ncbi.nlm.nih.gov/pubmed/9485051.

Kafitz KW, Greer CA. 1999. Olfactory ensheathing cells promote neurite extension from embryonic olfactory receptor cells in vitro. Glia. 25(2):99–110. doi:10.1002/(SICI)1098-1136(19990115)25:2<99::AID-GLIA1>3.0.CO;2-V.

Kaplan MS, Hinds JW. 1977. Neurogenesis in the adult rat: Electron microscopic analysis of light radioautographs. Science (80-). 197(4308):1092–1094. doi:10.1126/science.887941.

Kasowski HJ, Kim H, Greer CA. 1999. Compartmental organization of the olfactory bulb glomerulus. J Comp Neurol. 407(2):261–274. doi:10.1002/(SICI)1096-9861(19990503)407:2<261::AID-CNE7>3.0.CO;2-G.

Kawagishi K, Ando M, Yokouchi K, Sumitomo N, Karasawa M, Fukushima N, Moriizumi T. 2014. Stereological quantification of olfactory receptor neurons in mice. Neuroscience. 272:29–33. doi:10.1016/j.neuroscience.2014.04.050.

Kielian T. 2006. Toll-like receptors in central nervous system glial inflammation and homeostasis. J Neurosci Res. 83(5):711–730. doi:10.1002/jnr.20767. [accessed 2020 Jul 1]. /pmc/articles/PMC2440498/?report=abstract.

Kitayama M, Ueno M, Itakura T, Yamashita T. 2011. Activated microglia inhibit axonal growth through RGMa. PLoS One. 6(9). doi:10.1371/journal.pone.0025234. [accessed 2020 Jul 1]. https://pubmed.ncbi.nlm.nih.gov/21957482/.

Kobayashi M, Costanzo RM. 2009. Olfactory Nerve Recovery Following Mild and Severe Injury and the Efficacy of Dexamethasone Treatment. Chem Senses. 34(7):573–580. doi:10.1093/chemse/bjp038. [accessed 2019 Sep 16]. https://academic.oup.com/chemse/article-lookup/doi/10.1093/chemse/bjp038.

Koizumi S, Shigemoto-Mogami Y, Nasu-Tada K, Shinozaki Y, Ohsawa K, Tsuda M, Joshi B V., Jacobson KA, Kohsaka S, Inoue K. 2007. UDP acting at P2Y6 receptors is a mediator of microglial phagocytosis. Nature. 446(7139):1091–1095. doi:10.1038/nature05704.

Kuhn HG, Dickinson-Anson H, Gage FH. 1996. Neurogenesis in the dentate gyrus of the adult rat: Age-related decrease of neuronal progenitor proliferation. J Neurosci. 16(6):2027–2033. doi:10.1523/jneurosci.16-06-02027.1996.

Largent BL, Sosnowski RG, Reed RR. 1993. Directed expression of an oncogene to the olfactory neuronal lineage in transgenic mice. J Neurosci. 13(1):300–312. doi:10.1523/jneurosci.13-01-00300.1993.

Leonardi-essmann F, Emig M, Kitamura Y, Spanagel R, Gebicke-haerter PJ. 2005. Fractalkine-upregulated milk-fat globule EGF factor-8 protein in cultured rat microglia. J Neuroimmunol. 160:92–101. doi:10.1016/j.jneuroim.2004.11.012.

Leung CT, Coulombe PA, Reed RR. 2007. Contribution of olfactory neural stem cells to tissue maintenance and regeneration. Nat Neurosci. 10(6):720–726. doi:10.1038/nn1882.

Levai O, Breer H, Strotmann J. 2003. Subzonal organization of olfactory sensory neurons projecting to distinct glomeruli within the mouse olfactory bulb. J Comp Neurol. 458(3):209–220. doi:10.1002/cne.10559.

Li Y, Wang J, Sheng JG, Liu L, Barger SW, Jones RA, Van Eldik LJ, Mrak RE, Griffin WST. 1998. S100β Increases Levels of β-Amyloid Precursor Protein and Its Encoding mRNA in Rat Neuronal Cultures. J Neurochem. 71(4):1421–1428. doi:10.1046/j.1471-4159.1998.71041421.x. [accessed 2020 Feb 15]. http://doi.wiley.com/10.1046/j.1471-4159.1998.71041421.x.

Liberia T, Martin-Lopez E, Meller SJ, Greer CA. 2019. Sequential maturation of olfactory sensory neurons in the mature olfactory epithelium. eNeuro. 6(5):ENEURO.0266-19.2019. doi:10.1523/eneuro.0266-19.2019.

Ling G, Gu J, Genter MB, Zhuo X, Ding X. 2004. Regulation of cytochrome P450 gene expression in the olfactory mucosa. Chem Biol Interact. 147:247–258. doi:10.1016/j.cbi.2004.02.003.

Linheng L, Clevers H. 2010. Coexistence of quiescent and active adult stem cells in mammals. Science (80-). 327(5965):542–545. doi:10.1126/science.1180794.

Linnartz B, Kopatz J, Tenner AJ, Neumann H. 2012. Sialic acid on the neuronal glycocalyx prevents complement c1 binding and complement receptor-3-mediated removal by microglia. J Neurosci. 32(3):946–952. doi:10.1523/JNEUROSCI.3830-11.2012.

Lipscomb B, Treloar H, Greer C. 2002. Cell surface carbohydrates reveal heterogeneity in olfactory receptor cell axons in the mouse. Cell Tissue Res. 308(1):7–17. doi:10.1007/s00441-002-0532-0. [accessed 2019 Sep 9]. http://link.springer.com/10.1007/s00441-002-0532-0.

Lipscomb BW, Treloar HB, Greer CA. 2002. Novel microglomerular structures in the olfactory bulb of mice. J Neurosci. 22(3):766–74. doi:10.1523/JNEUROSCI.22-

03-00766.2002. [accessed 2019 Sep 9].
http://www.ncbi.nlm.nih.gov/pubmed/11826106.

Lukácsi S, Nagy-Baló Z, Erdei A, Sándor N, Bajtay Z. 2017. The role of CR3
(CD11b/CD18) and CR4 (CD11c/CD18) in complement-mediated phagocytosis
and podosome formation by human phagocytes. Immunol Lett. 189:64–72.
doi:10.1016/j.imlet.2017.05.014.

Luskin MB. 1993. Restricted proliferation and migration of postnatally generated
neurons derived from the forebrain subventricular zone. Neuron. 11(1):173–189.
doi:10.1016/0896-6273(93)90281-U.

Ma M. 2010. Multiple Olfactory Subsystems Convey Various Sensory Signals. CRC
Press/Taylor & Francis. [accessed 2019 Aug 19].
http://www.ncbi.nlm.nih.gov/pubmed/21882425.

Mackay-Sim A, Kittel P. 1991. Cell dynamics in the adult mouse olfactory epithelium: A
quantitative autoradiographic study. J Neurosci. 11(4):979–984.
doi:10.1523/jneurosci.11-04-00979.1991.

Marillat V, Cases O, Nguyenf-Ba-Charvet KT, Tessier-Lavigne M, Sotelo C, Chédotal A.
2002. Spatiotemporal expression patterns of slit and robo genes in the rat brain.
J Comp Neurol. 442(2):130–155. doi:10.1002/cne.10068. [accessed 2019 Sep
7]. http://doi.wiley.com/10.1002/cne.10068.

Mayeur A, Duclos C, Honoré A, Gauberti M, Drouot L, do Rego JC, Bon-Mardion N,
Jean L, Vérin E, Emery E, et al. 2013. Potential of Olfactory Ensheathing Cells
from Different Sources for Spinal Cord Repair. PLoS One. 8(4).
doi:10.1371/journal.pone.0062860.

McIntyre JC, Titlow WB, McClintock TS. 2010. Axon growth and guidance genes
identify nascent, immature, and mature olfactory sensory neurons. J Neurosci
Res. 88(15):3243–3256. doi:10.1002/jnr.22497. [accessed 2019 Sep 16].
http://doi.wiley.com/10.1002/jnr.22497.

Mckimmie CS, Moore M, Fraser AR, Jamieson T, Xu D, Burt C, Pitman NI, Nibbs RJ,
Mcinnes IB, Liew FY, et al. 2009. A TLR2 ligand suppresses inflammation by
modulation of chemokine receptors and redirection of leukocyte migration. Blood.
113(18):4224–4232. doi:10.1182/blood-2008-08-174698.The.

Milde R, Ritter J, Tennent GA, Loesch A, Martinez FO, Gordon S, Pepys MB, Verschoor
A, Helming L. 2015. Multinucleated Giant Cells Are Specialized for Complement-
Mediated Phagocytosis and Large Target Destruction. Cell Rep. 13(9):1937–
1948. doi:10.1016/j.celrep.2015.10.065. [accessed 2020 Jun 30].
/pmc/articles/PMC4675895/?report=abstract.

Miragall F, Kadmon G, Husmann M, Schachner M. 1988. Expression of cell adhesion
molecules in the olfactory system of the adult mouse: Presence of the embryonic
form of N-CAM. Dev Biol. 129(2):516–531. doi:10.1016/0012-1606(88)90397-1.

Miragall F, Kadmon G, Schachner M. 1989. Expression of L1 and N-CAM cell adhesion molecules during development of the mouse olfactory system. Dev Biol. 135(2):272–286. doi:10.1016/0012-1606(89)90179-6.

Miyamichi K, Serizawa S, Kimura HM, Sakano H. 2005. Continuous and Overlapping Expression Domains of Odorant Receptor Genes in the Olfactory Epithelium Determine the Dorsal/Ventral Positioning ofGlomeruli in the Olfactory Bulb. J Neurosci. 25(14):3586–3592. doi:https://doi.org/10.1523/JNEUROSCI.0324-05.2005. [accessed 2019 Sep 7]. http://www.ncbi.nlm.nih.gov/pubmed/11943806.

Mohrhardt J, Nagel M, Fleck D, Ben-Shaul Y, Spehr M. 2018. Signal detection and coding in the accessory olfactory system. Chem Senses. 43(9):667–695. doi:10.1093/chemse/bjy061.

Mombaerts P. 2006. Axonal Wiring in the Mouse Olfactory System. Annu Rev Cell Dev Biol. 22(1):713–737. doi:10.1146/annurev.cellbio.21.012804.093915. [accessed 2019 Sep 9]. http://www.annualreviews.org/doi/10.1146/annurev.cellbio.21.012804.093915.

Morales-Medina JC, Iannitti T, Freeman A, Caldwell HK. 2017. The olfactory bulbectomized rat as a model of depression: The hippocampal pathway. Behav Brain Res. 317:562–575. doi:10.1016/j.bbr.2016.09.029.

Moreno-Flores MT, Díaz-Nido J, Wandosell F, Avila J. 2002. Olfactory ensheathing glia: Drivers of axonal regeneration in the central nervous system? J Biomed Biotechnol. 2(1):37–43. doi:10.1155/S1110724302000372.

Moreno-Flores MT, Lim F, Martín-Bermejo MJ, Díaz-Nido J, Ávila J, Wandosell F. 2003. High level of amyloid precursor protein expression in neurite-promoting olfactory ensheathing glia (OEG) and OEG-derived cell lines. J Neurosci Res. 71(6):871–881. doi:10.1002/jnr.10527. [accessed 2020 Feb 11]. http://doi.wiley.com/10.1002/jnr.10527.

Morrison EE, Costanzo RM. 1995. Regeneration of olfactory sensory neurons and reconnection in the aging hamster central nervous system. Neurosci Lett. 198(3):213–217. doi:10.1016/0304-3940(95)11943-Q.

Murdoch B, Roskams AJ. 2007. Olfactory epithelium progenitors: insights from transgenic mice and in vitro biology. J Mol Hist. 38(6):581–599. doi:10.1007/s10735-007-9141-2.

Musashe DT, Purice MD, Speese SD, Doherty J, Logan MA. 2016. Insulin-like Signaling Promotes Glial Phagocytic Clearance of Degenerating Axons through Regulation of Draper. Cell Rep. 16(7):1838–1850. doi:10.1016/j.celrep.2016.07.022. [accessed 2020 Feb 14]. http://dx.doi.org/10.1016/j.celrep.2016.07.022.

Nakashima A, Takeuchi H, Imai T, Saito H, Kiyonari H, Abe T, Chen M, Weinstein LS, Yu CR, Storm DR, et al. 2013. Agonist-independent GPCR activity regulates anterior-posterior targeting of olfactory sensory neurons. Cell. 154(6):1314–1325. doi:10.1016/j.cell.2013.08.033.

Nazareth L, Lineburg KE, Chuah MI, Tello Velasquez J, Chehrehasa F, St John JA,

Ekberg JAK. 2015. Olfactory ensheathing cells are the main phagocytic cells that remove axon debris during early development of the olfactory system. J Comp Neurol. 523(3):479–494. doi:10.1002/cne.23694.

Neumann H, Kotter MR, Franklin RJM. 2009. Debris clearance by microglia: An essential link between degeneration and regeneration. Brain. 132(2):288–295. doi:10.1093/brain/awn109.

Nimmerjahn A, Kirchhoff F, Helmchen F. 2005. Resting microglial cells are highly dynamic surveillants of brain parenchyma in vivo. Science (80-). 308(5726):1314–1318. doi:10.1126/science.1110647.

Noda M, Doi Y, Liang J, Kawanokuchi J, Sonobe Y, Takeuchi H, Mizuno T, Suzumura A. 2011. Fractalkine attenuates excito-neurotoxicity via microglial clearance of damaged neurons and antioxidant enzyme heme oxygenase-1 expression. J Biol Chem. 286(3):2308–2319. doi:10.1074/jbc.M110.169839.

Norlin EM, Alenius M, Gussing F, Hägglund M, Vedin V, Bohm S. 2001. Evidence for gradients of gene expression correlating with zonal topography of the olfactory sensory map. Mol Cell Neurosci. 18(3):283–295. doi:10.1006/mcne.2001.1019.

Ogawa T, Takezawa K, Shimizu S, Shimizu T. 2014. Valproic acid promotes neural regeneration of olfactory epithelium in adult mice after methimazole-induced damage. Am J Rhinol Allergy. 28(2):95–99. doi:10.2500/ajra.2014.28.4027.

Pays L, Schwarting G. 2000. Gal-NCAM is a differentially expressed marker for mature sensory neurons in the rat olfactory system. J Neurobiol. 43(2):173–85. [accessed 2019 Sep 9]. http://www.ncbi.nlm.nih.gov/pubmed/10770846.

Pedersen PE, Jastreboff PJ, Stewart WB, Shepherd GM. 1986. Mapping of an olfactory receptor population that projects to a specific region in the rat olfactory bulb. J Comp Neurol. 250(1):93–108. doi:10.1002/cne.902500109. [accessed 2020 Feb 1]. http://doi.wiley.com/10.1002/cne.902500109.

Peri F, Nüsslein-Volhard C. 2008. Live Imaging of Neuronal Degradation by Microglia Reveals a Role for v0-ATPase a1 in Phagosomal Fusion In Vivo. Cell. 133(5):916–927. doi:10.1016/j.cell.2008.04.037.

Pevny L, Rao MS. 2003. The stem-cell menagerie. Trends Neurosci. 26(7):351–359. doi:10.1016/S0166-2236(03)00169-3.

Pixley SK. 1992. The olfactory nerve contains two populations of glia, identified both in vivo and in vitro. Glia. 5(4):269–284. doi:10.1002/glia.440050405.

Pont-Lezica L, Beumer W, Colasse S, Drexhage H, Versnel M, Bessis A. 2014. Microglia shape corpus callosum axon tract fasciculation: functional impact of prenatal inflammation. Eur J Neurosci. 39(10):1551–1557. doi:10.1111/ejn.12508. [accessed 2020 Apr 1]. http://doi.wiley.com/10.1111/ejn.12508.

Ramón-Cueto A, Avila J. 1998. Olfactory ensheathing glia: Properties and function. Brain Res Bull. 46(3):175–187. doi:10.1016/S0361-9230(97)00463-2.

Reemst K, Noctor SC, Lucassen PJ, Hol EM. 2016. The indispensable roles of microglia and astrocytes during brain development. Front Hum Neurosci. 10(NOV2016). doi:10.3389/fnhum.2016.00566.

Reshef R, Kudryavitskaya E, Shani-Narkiss H, Isaacson B, Rimmerman N, Mizrahi A, Yirmiya R. 2017. The role of microglia and their CX3CR1 signaling in adult neurogenesis in the olfactory bulb. Elife. 6. doi:10.7554/eLife.30809.

Ressler KJ, Sullivan SL, Buck LB. 1993. A zonal organization of odorant receptor gene expression in the olfactory epithelium. Cell. 73(3):597–609. doi:10.1016/0092-8674(93)90145-g. [accessed 2019 Sep 7]. http://www.ncbi.nlm.nih.gov/pubmed/7683976.

Richard MB, Taylor SR, Greer CA. 2010. Age-induced disruption of selective olfactory bulb synaptic circuits. Proc Natl Acad Sci U S A. 107(35):15613–8. doi:10.1073/pnas.1007931107. [accessed 2019 Sep 7]. http://www.ncbi.nlm.nih.gov/pubmed/20679234.

Ring G, Mezza RC, Schwob JE. 1997. Immunohistochemical identification of discrete subsets of rat olfactory neurons and the glomeruli that they innervate. J Comp Neurol. 388(3):415–434. doi:10.1002/(SICI)1096-9861(19971124)388:3<415::AID-CNE5>3.0.CO;2-3. [accessed 2019 Sep 9]. http://doi.wiley.com/10.1002/%28SICI%291096-9861%2819971124%29388%3A3%3C415%3A%3AAID-CNE5%3E3.0.CO%3B2-3.

Robinson AM, Conley DS, Shinners MJ, Kern RC. 2002. Apoptosis in the aging olfactory epithelium. Laryngoscope. 112(8):1431–1435. doi:10.1097/00005537-200208000-00019.

Rodriguez-Gil DJ, Bartel DL, Jaspers AW, Mobley AS, Imamura F, Greer C a. 2015. Odorant receptors regulate the final glomerular coalescence of olfactory sensory neuron axons. Proc Natl Acad Sci.:201417955. doi:10.1073/pnas.1417955112.

Rodriguez-Gil DJ, Greer CA. 2008. Wnt/frizzled family members mediate olfactory sensory neuron axon extension. J Comp Neurol. 511(3):301–317. doi:10.1002/cne.21834. [accessed 2019 Sep 7]. http://doi.wiley.com/10.1002/cne.21834.

Sakai M, Nagatsu I. 1993. Alteration of carnosine expression in olfactory system of mouse after unilateral naris closure and partial bulbectomy. J Neurosci Res. 34(6):648–653. doi:10.1002/jnr.490340608. [accessed 2019 Sep 16]. http://doi.wiley.com/10.1002/jnr.490340608.

Satoda M, Takagi S, Ohta K, Hirata T, Fujisawa H. 1995. Differential expression of two cell surface proteins, neuropilin and plexin, in Xenopus olfactory axon subclasses. J Neurosci. 15(1 II):942–955. doi:10.1523/jneurosci.15-01-00942.1995.

Schafer DP, Lehrman EK, Kautzman AG, Koyama R, Mardinly AR, Yamasaki R, Ransohoff RM, Greenberg ME, Barres BA, Stevens B. 2012. Microglia Sculpt

Postnatal Neural Circuits in an Activity and Complement-Dependent Manner. Neuron. 74(4):691–705. doi:10.1016/j.neuron.2012.03.026.

Schain M, Kreisl WC. 2017. Neuroinflammation in Neurodegenerative Disorders—a Review. Curr Neurol Neurosci Rep. 17(3). doi:10.1007/s11910-017-0733-2.

Schoenfeld TA, Clancy AN, Forbes WB, Macrides F. 1994. The spatial organization of the peripheral olfactory system of the hamster. part I: Receptor neuron projections to the main olfactory bulb. Brain Res Bull. 34(3):183–210. doi:10.1016/0361-9230(94)90059-0.

Schwarting GA, Henion TR. 2008. Olfactory axon guidance: The modified rules. J Neurosci Res. 86(1):11–17. doi:10.1002/jnr.21373. [accessed 2019 Sep 7]. http://doi.wiley.com/10.1002/jnr.21373.

Schwob JE. 2002. Neural regeneration and the peripheral olfactory system. Anat Rec. 269(1):33–49. doi:10.1002/ar.10047. [accessed 2020 Apr 15]. http://doi.wiley.com/10.1002/ar.10047.

Schwob JE, Costanzo RM. 2010. Regeneration of the Olfactory Epithelium. Senses A Compr Ref. 4:591–612. doi:10.1016/B978-012370880-9.00115-8.

Schwob JE, Youngentob SL, Mezza RC. 1995. Reconstitution of the rat olfactory epithelium after methyl bromide-induced lesion. J Comp Neurol. 359(1):15–37. doi:10.1002/cne.903590103. [accessed 2019 Sep 16]. http://doi.wiley.com/10.1002/cne.903590103.

Schwob JE, Youngentob SL, Ring G, Iwema CL, Mezza RC. 1999. Reinnervation of the rat olfactory bulb after methyl bromide-induced lesion: Timing and extent of reinnervation. J Comp Neurol. 412(3):439–457. doi:10.1002/(SICI)1096-9861(19990927)412:3<439::AID-CNE5>3.0.CO;2-H. [accessed 2019 Sep 16]. http://doi.wiley.com/10.1002/%28SICI%291096-9861%2819990927%29412%3A3%3C439%3A%3AAID-CNE5%3E3.0.CO%3B2-H.

Seki T, Arai Y. 1993. Highly polysialylated neural cell adhesion molecule (NCAM-H) is expressed by newly generated granule cells in the dentate gyrus of the adult rat. J Neurosci. 13(6):2351–2358. doi:10.1523/jneurosci.13-06-02351.1993.

Serizawa S, Miyamichi K, Nakatani H, Suzuki M, Saito M, Yoshihara Y, Sakano H. 2003. Negative feedback regulation ensures the one receptor-one olfactory neuron rule in mouse. Science. 302(5653):2088–94. doi:10.1126/science.1089122. [accessed 2019 Sep 7]. http://www.ncbi.nlm.nih.gov/pubmed/14593185.

Sierra A, Abiega O, Shahraz A, Neumann H. 2013. Janus-faced microglia: Beneficial and detrimental consequences of microglial phagocytosis. Front Cell Neurosci. 7(6). doi:10.3389/fncel.2013.00006.

Silverman SM, Kim BJ, Howell GR, Miller J, John SWM, Wordinger RJ, Clark AF. 2016. C1q propagates microglial activation and neurodegeneration in the visual axis following retinal ischemia/reperfusion injury. Mol Neurodegener. 11(1). doi:10.1186/s13024-016-0089-0.

Slotnick B, Cockerham R, Pickett E. 2004. Olfaction in olfactory bulbectomized rats. J Neurosci. 24(41):9195–9200. doi:10.1523/JNEUROSCI.1936-04.2004.

Smith SG, Northcutt K V. 2018. Perinatal hypothyroidism increases play behaviors in juvenile rats. Horm Behav. 98:1–7. doi:10.1016/j.yhbeh.2017.11.012.

Smithson LJ, Kawaja MD. 2010. Microglial/macrophage cells in mammalian olfactory nerve fascicles. J Neurosci Res. 88(4):858–865. doi:10.1002/jnr.22254.

Song C, Leonard BE. 2005. The olfactory bulbectomised rat as a model of depression. Neurosci Biobehav Rev. 29(4–5):627–647. doi:10.1016/j.neubiorev.2005.03.010.

Squarzoni P, Oller G, Hoeffel G, Pont-Lezica L, Rostaing P, Low D, Bessis A, Ginhoux F, Garel S. 2014. Microglia Modulate Wiring of the Embryonic Forebrain. Cell Rep. 8(5):1271–1279. doi:10.1016/j.celrep.2014.07.042.

St John JA, Claxton C, Robinson MW, Yamamoto F, Domino SE, Key B. 2006. Genetic manipulation of blood group carbohydrates alters development and pathfinding of primary sensory axons of the olfactory systems. Dev Biol. 298(2):470–484. doi:10.1016/J.YDBIO.2006.06.052. [accessed 2019 Sep 9]. https://www.sciencedirect.com/science/article/pii/S0012160606009778?via%3Dih ub.

St John JA, Key B. 2001. EphB2 and two of its ligands have dynamic protein expression patterns in the developing olfactory system. Dev Brain Res. 126(1):43–56. doi:10.1016/S0165-3806(00)00136-X. [accessed 2019 Sep 9]. https://www.sciencedirect.com/science/article/pii/S016538060000136X?via%3Di hub.

Van Strijp JA, Russell DG, Tuomanen E, Brown EJ, Wright SD. 1993. Ligand specificity of purified complement receptor type three (CD11b/CD18, alpha m beta 2, Mac-1). Indirect effects of an Arg-Gly-Asp (RGD) sequence. J Immunol. 151(6).

Strotmann J, Conzelmann S, Beck A, Feinstein P, Breer H, Mombaerts P. 2000. Local permutations in the glomerular array of the mouse olfactory bulb. J Neurosci. 20(18):6927–38. doi:10.1523/JNEUROSCI.20-18-06927.2000. [accessed 2019 Sep 7]. http://www.ncbi.nlm.nih.gov/pubmed/10995837.

Su Z, Chen J, Qiu Y, Yuan Y, Zhu F, Zhu Y, Liu X, Pu Y, He C. 2013. Olfactory ensheathing cells: The primary innate immunocytes in the olfactory pathway to engulf apoptotic olfactory nerve debris. Glia. 61(4):490–503. doi:10.1002/glia.22450.

Sullivan CS, Scheib JL, Ma Z, Dang RP, Schafer JM, Hickman FE, Brodsky FM, Ravichandran KS, Carter BD. 2014. The adaptor protein GULP promotes Jedi-1-mediated phagocytosis through a clathrin-dependent mechanism. Mol Biol Cell. 25(12):1925–1936. doi:10.1091/mbc.E13-11-0658.

Suzukawa K, Kondo K, Kanaya K, Sakamoto T, Watanabe K, Ushio M, Kaga K,

Yamasoba T. 2011. Age-related changes of the regeneration mode in the mouse peripheral olfactory system following olfactotoxic drug methimazole-induced damage. J Comp Neurol. 519(11):2154–2174. doi:10.1002/cne.22611.

Suzuki Y, Takeda M, Farbman AI. 1996. Supporting cells as phagocytes in the olfactory epithelium after bulbectomy. J Comp Neurol. 376(4):509–517. doi:10.1002/(SICI)1096-9861(19961223)376:4<509::AID-CNE1>3.0.CO;2-5.

Takahashi H, Yoshihara S ichi, Nishizumi H, Tsuboi A. 2010. Neuropilin-2 is required for the proper targeting of ventral glomeruli in the mouse olfactory bulb. Mol Cell Neurosci. 44(3):233–245. doi:10.1016/j.mcn.2010.03.010.

Takeuchi H, Inokuchi K, Aoki M, Suto F, Tsuboi A, Matsuda I, Suzuki M, Aiba A, Serizawa S, Yoshihara Y, et al. 2010. Sequential arrival and graded secretion of Sema3F by olfactory neuron axons specify map topography at the bulb. Cell. 141(6):1056–1067. doi:10.1016/j.cell.2010.04.041.

Takeuchi H, Sakano H. 2014. Neural map formation in the mouse olfactory system. Cell Mol Life Sci. 71(16):3049–3057. doi:10.1007/s00018-014-1597-0.

Treloar HB, Bartolomei JC, Lipscomb BW, Greer CA. 2001. Mechanisms of Axonal Plasticity: Lessons from the Olfactory Pathway. Neurosci. 7(1):55–63. doi:10.1177/107385840100700109. [accessed 2019 Sep 7]. http://journals.sagepub.com/doi/10.1177/107385840100700109.

Treloar HB, Feinstein P, Mombaerts P, Greer CA. 2002. Specificity of glomerular targeting by olfactory sensory axons. J Neurosci. 22(7):2469–77. doi:20026239. [accessed 2019 Sep 7]. http://www.ncbi.nlm.nih.gov/pubmed/11923411.

Ubink R, Halasz N, Zhang X, Dagerlind Å, Hökfelt T. 1994. Neuropeptide tyrosine is expressed in ensheathing cells around the olfactory nerves in the rat olfactory bulb. Neuroscience. 60(3):709–726. doi:10.1016/0306-4522(94)90499-5.

Ubink R, Hökfelt T. 2000. Expression of neuropeptide Y in olfactory ensheathing cells during prenatal development. J Comp Neurol. 423(1):13–25. doi:10.1002/1096-9861(20000717)423:1<13::AID-CNE2>3.0.CO;2-P. [accessed 2020 Feb 15]. http://doi.wiley.com/10.1002/1096-9861%2820000717%29423%3A1%3C13%3A%3AAID-CNE2%3E3.0.CO%3B2-P.

Vassar R, Ngai J, Axel R. 1993. Spatial segregation of odorant receptor expression in the mammalian olfactory epithelium. Cell. 74(2):309–318. doi:10.1016/0092-8674(93)90422-M.

Vilalta A, Brown GC. 2018. Neurophagy, the phagocytosis of live neurons and synapses by glia, contributes to brain development and disease. FEBS J. 285(19):3566–3575. doi:10.1111/febs.14323.

Vincent AJ, Taylor JM, Choi-Lundberg DL, West AK, Chuah MI. 2005. Genetic expression profile of olfactory ensheathing cells is distinct from that of Schwann cells and astrocytes. Glia. 51(2):132–147. doi:10.1002/glia.20195. [accessed

2020 Feb 15]. http://doi.wiley.com/10.1002/glia.20195.

Vincent AJ, West AK, Meng IC. 2005. Morphological and functional plasticity of olfactory ensheathing cells. J Neurocytol. 34(1–2):65–80. doi:10.1007/s11068-005-5048-6.

Wallace J, Lord J, Dissing-Olesen L, Stevens B, Murthy VN. 2020. Microglial depletion disrupts normal functional development of adult-born neurons in the olfactory bulb. Elife. 9. doi:10.7554/eLife.50531.

Walz A, Rodriguez I, Mombaerts P. 2002. Aberrant sensory innervation of the olfactory bulb in neuropilin-2 mutant mice. J Neurosci. 22(10):4025–35. doi:20026374. [accessed 2019 Sep 9]. http://www.ncbi.nlm.nih.gov/pubmed/12019322.

Wang F, Nemes A, Mendelsohn M, Axel R. 1998. Odorant receptors govern the formation of a precise topographic map. Cell. 93(1):47–60. doi:10.1016/s0092-8674(00)81145-9. [accessed 2019 Sep 7]. http://www.ncbi.nlm.nih.gov/pubmed/9546391.

Wang Q, Liu Y, Zhou J. 2015. Neuroinflammation in Parkinson's disease and its potential as therapeutic target. Transl Neurodegener. 4(1). doi:10.1186/s40035-015-0042-0.

Wang Y-Z, Molotkov A, Song L, Li Y, Pleasure DE, Zhou C-J. 2008. Activation of the Wnt/β-catenin signaling reporter in developing mouse olfactory nerve layer marks a specialized subgroup of olfactory ensheathing cells. Dev Dyn. 237(11):3157–3168. doi:10.1002/dvdy.21712. [accessed 2019 Sep 7]. http://doi.wiley.com/10.1002/dvdy.21712.

Weinger JG, Brosnan CF, Loudig O, Goldberg MF, Macian F, Arnett HA, Prieto AL, Tsiperson V, Shafit-Zagardo B. 2011. Loss of the receptor tyrosine kinase Axl leads to enhanced inflammation in the CNS and delayed removal of myelin debris during Experimental Autoimmune Encephalomyelitis. J Neuroinflammation. 8:49. doi:10.1186/1742-2094-8-49. [accessed 2020 Jul 1]. /pmc/articles/PMC3121615/?report=abstract.

Whitby-Logan GK, Weech M, Walters E. 2004. Zonal expression and activity of glutathione S-transferase enzymes in the mouse olfactory mucosa. Brain Res. 995:151–157. doi:10.1016/j.brainres.2003.09.012.

White DM. 1998. Contribution of neurotrophin-3 to the neuropeptide Y-induced increase in neurite outgrowth of rat dorsal root ganglion cells. Neuroscience. 86(1):257–263. doi:10.1016/S0306-4522(98)00034-7.

White DM, Mansfield K. 1996. Vasoactive intestinal polypeptide and neuropeptide Y act indirectly to increase neurite outgrowth of dissociated dorsal root ganglion cells. Neuroscience. 73(3):881–887. doi:10.1016/0306-4522(96)00055-3.

Whitman MC, Greer CA. 2009. Adult neurogenesis and the olfactory system. Prog Neurobiol. doi:10.1016/j.pneurobio.2009.07.003.

Williams-Hogarth LC, Puche AC, Torrey C, Cai X, Song I, Kolodkin AL, Shipley MT, Ronnett G V. 2000. Expression of semaphorins in developing and regenerating olfactory epithelium. J Comp Neurol. 423(4):565–578. doi:10.1002/1096-9861(20000807)423:4<565::AID-CNE3>3.0.CO;2-F. [accessed 2019 Sep 9]. http://doi.wiley.com/10.1002/1096-9861%2820000807%29423%3A4%3C565%3A%3AAID-CNE3%3E3.0.CO%3B2-F.

Williams SK, Gilbey T, Barnett SC. 2004. Immunohistochemical Studies of the Cellular Changes in the Peripheral Olfactory System After Zinc Sulfate Nasal Irrigation. Neurochem Res. 29(5):891–901. doi:10.1023/B:NERE.0000021234.46315.34. [accessed 2019 Sep 16]. http://link.springer.com/10.1023/B:NERE.0000021234.46315.34.

Wu HH, Bellmunt E, Scheib JL, Venegas V, Burkert C, Reichardt LF, Zhou Z, Farinas I, Carter BD, Farías I, et al. 2009. Glial precursors clear sensory neuron corpses during development via Jedi-1, an engulfment receptor. Nat Neurosci. 12(12):1534–1541. doi:10.1038/nn.2446.Glial.

Yee KK, Costanzo RM. 1995. Restoration of olfactory mediated behavior after olfactory bulb deafferentation. Physiol Behav. 58(5):959–968. doi:10.1016/0031-9384(95)00159-G.

Yee KK, Costanzo RM. 1998. Changes in odor quality discrimination following recovery from olfactory nerve transection. Chem Senses. 23(5):513–519. doi:10.1093/chemse/23.5.513.

Yoshihara Y, Kawasaki M, Tamada A, Fujita H, Hayashi H, Kagamiyama H, Mori K. 1997. OCAM: A new member of the neural cell adhesion molecule family related to zone-to-zone projection of olfactory and vomeronasal axons. J Neurosci. 17(15):5830–5842. doi:10.1523/jneurosci.17-15-05830.1997.

Yoshiyama Y, Higuchi M, Zhang B, Huang SM, Iwata N, Saido TC, Maeda J, Suhara T, Trojanowski JQ, Lee VMY. 2007. Synapse Loss and Microglial Activation Precede Tangles in a P301S Tauopathy Mouse Model. Neuron. 53(3):337–351. doi:10.1016/j.neuron.2007.01.010.

Yu CR, Wu Y. 2017. Regeneration and rewiring of rodent olfactory sensory neurons. Exp Neurol. 287:395–408. doi:10.1016/j.expneurol.2016.06.001.

Zaghetto AA, Paina S, Mantero S, Platonova N, Peretto P, Bovetti S, Puche A, Piccolo S, Merlo GR. 2007. Activation of the Wnt-beta catenin pathway in a cell population on the surface of the forebrain is essential for the establishment of olfactory axon connections. J Neurosci. 27(36):9757–68. doi:10.1523/JNEUROSCI.0763-07.2007. [accessed 2019 Sep 9]. http://www.ncbi.nlm.nih.gov/pubmed/17804636.

Zamparo I, Francia S, Franchi SA, Redolfi N, Costanzi E, Kerstens A, Fukutani Y, Battistutta R, Polverino de Laureto P, Munck S, et al. 2019. Axonal Odorant Receptors Mediate Axon Targeting. Cell Rep. doi:10.1016/j.celrep.2019.11.099.

Zen K, Guo YL, Li LM, Bian Z, Zhang CY, Liu Y. 2011. Cleavage of the CD11b extracellular domain by the leukocyte serprocidins is critical for neutrophil

detachment during chemotaxis. Blood. 117(18):4885–4894. doi:10.1182/blood-2010-05-287722. [accessed 2020 Jul 1].
/pmc/articles/PMC3100697/?report=abstract.

www.ingramcontent.com/pod-product-compliance
Lightning Source LLC
LaVergne TN
LVHW041729190726
843493LV00007B/2277